BIOLOGICAL
INVESTIGATIONS
IN THE
LABORATORY

BIOLOGICAL
INVESTIGATIONS
IN THE
LABORATORY

WILLIAM T. KEETON

MICHAEL W. DABNEY

MARY J. PHILPOTT

PRINCETON UNIVERSITY

W • W • NORTON & COMPANY • NEW YORK • LONDON

The text of this book is composed in Press Roman. Composition and layout by
Roberta Flechner.

ISBN 0-393-95260-6

W. W. Norton & Company, Inc., 500 Fifth Avenue, New York, N.Y. 10110
W. W. Norton & Company Ltd., 37 Great Russell Street, London WC1B 3NU

1 2 3 4 5 6 7 8 9 0

CONTENTS

NOTE TO
THE INSTRUCTOR

Biological Investigations in the Laboratory represents an effort to integrate observational and experimental biology and to provide introductory students with essential laboratory experience. Its goals are (1) to provide an opportunity for seeing and handling a wide variety of biological materials (alive, whenever possible) that many students might not otherwise encounter; (2) to develop a feel for the design and limitations of experiments as well as the excitement of independent discovery; (3) to promote an appreciation for the extraordinary variety and complexity of living organisms; (4) to develop proficiency in the use of standard laboratory equipment and in the proper handling of materials; and (5) to foster the development of self-sufficiency in the conduct of laboratory experiments and observations, and in the analysis of data.

Although this manual has been tailored to accompany *Biological Science,* fourth edition, by William T. Keeton and James L. Gould, and *Elements of Biological Science,* third edition, by William T. Keeton and Carol Hardy McFadden, it will complement most introductory biology texts and is adaptable to most introductory biology courses. Since the manual contains more exercises than can be completed in an academic year, the instructor may present those Topics best suited to the needs of the class while reserving others for independent research projects that permit the student to investigate a subject in greater detail.

As a revision of *Laboratory Guide for Biological Science,* this manual represents an effort to retain the best aspects of its predecessor, to clarify those sections that have presented problems, and to update all facts and biological terminology. Every Topic has been carefully reviewed for accuracy and improved methods, and each Topic has been read alongside corresponding materials in *Biological Science,* fourth edition. The Topics dealing with systematics have been updated to reflect the latest in classification and phylogenetic relationships. Like before, an evolutionary theme is stressed throughout *Biological Investigations.* Great care has been taken to make laboratory procedures clear to the student, thereby allowing the instructor more time to provide other valuable assistance to students in lab.

Because of the expense of laboratory materials and equipment, I have chosen inexpensive equipment when possible and have described procedures insuring optimal use of these materials. The *Instructor's Manual to Accompany Laboratory Guide for Biological Science* includes suggestions for equipment substitutions only when limited funds or materials are available. Since laboratory preparation is crucial and time-consuming, detailed equipment lists, sources of materials, updated costs, and suggestions for preparation are included in the *Instructor's Manual.*

Finally, it is a pleasure to acknowledge all those whose comments and criticism were invaluable in the preparation of this revision, particularly Carol Hardy McFadden of Cornell University. I also owe special thanks to Linda Cholewiak for her insights, comments, and suggestions, and to the several suppliers of biological materials and equipment who provided illustrations and materials for use in this guide.

I would also like to acknowledge the patience and assistance of the staff of W. W. Norton & Company, especially Jim Jordan.

Any comments or suggestions from users, faculty, and students are most welcome.

Princeton, New Jersey   M.J.P.
August 1986

NOTE TO
THE STUDENT

We live in a civilization dominated by science and its products. Whether one accepts the role of science in today's world or not, it is crucial for every educated individual to try to understand the goals and methods of science. As you will soon discover, one of the most important and interesting places to begin understanding science is in the laboratory, since it is here that the daily business of science often takes place.

Science is not a spectator sport. If you are to know it well, you must deal with it intimately. The purpose of laboratory experience, then, is to clarify the meaning and methods of science by giving you the opportunity to become a part-time scientist. Though it might not teach you to become a professional biologist, as experience it can provide background that will be important if you embark upon a career in the natural sciences. With your effort, this book and your instructor together can help you to learn to think effectively about biological problems; they will afford you firsthand contact with a wide variety of living organisms and a wide variety of biological ideas, and they will encourage you to do some biological exploring on your own and occasionally to carry out experiments of your own design.

Attention to the following suggestions will save you time and trouble:

1. Always *preread* the laboratory work, and come to your laboratory session with a clear idea of what it is you are to do and in what order. Be prepared to ask questions about anything you do not understand. Prereading will help you know what to bring to the laboratory and will clarify your instructor's supplementary directions. During the sessions, you will have little time to study the instructions and procedures, especially in laboratories where experiments are to be performed; you should be familiar with these beforehand.

2. One of the most important features of scientific inquiry is exchange of information. Scientists publish to make their results known to others. You should encourage the same kind of exchange with your neighbors and with your instructor. Remember that your laboratory instructor, whether graduate student or professor, is a professional biologist. While this does not mean that he or she has all the answers, your contacts with your instructor and the degree to which you profit from his or her experiences and insights form an important part of your laboratory experience.

3. This book and your instructor will pose many questions to encourage discussion and invite your thinking about biological problems. Become a questioner, too. Some of the questions we raise may be unanswerable either because they have not been tested or because by their nature they are not susceptible to experimental test. Try not to allow this to discourage you; rather, let it emphasize, as it does for many scientists, the open-ended and undogmatic nature of scientific inquiry.

4. Always know exactly what you are doing in the laboratory and why you are doing it. Nothing could be worse than to drift through a laboratory session in a state of confusion. While later laboratory sessions, readings, and lectures will clarify what has come before, you should always promptly clear up problems as you encounter them. If you feel you are losing perspective during a laboratory session, ask your instructor for help.

5. Consideration for others who use the laboratory facilities requires your cooperation in leaving the laboratory room as neat as you can—or, if possible, neater than you found it. Never wait for your instructor to ask you to clean up; take the initiative to clear your own area. Make sure that microscope lenses are perfectly clean and that the other equipment you use each week is properly stored.

6. Occasionally, demonstrations may be set up on one of the tables in the laboratory room. You are always responsible for the content of such demonstrations.

7. Bring your text to the laboratory. References to *Biological Science* (fourth edition) and *Elements of Biological Science* (third edition) are listed at the beginning of each Topic. If your class is using another text, your instructor will supply page references to it. Familiarize yourself with the location of additional reference books, and make liberal use of them.

8. Biology has a language of its own, which is succinct but which can be troublesome at times. Terms you should know are indicated throughout this book in **boldface**; some students find it helpful to maintain a list of such terms. Most of the scientific words you will encounter have Greek or Latin roots. If you learn these roots and use the words frequently, you will soon be comfortable even with unfamiliar terms in new contexts.

9. Most important of all, remember that this book is not meant to limit your laboratory experiences. It is only a guide; when it stops, you should continue. Experiment not just with what is required, but with anything and everything that interests you. And constantly observe the results.

We trust that your contact with science will stimulate more than a superficial curiosity about what it means to be a living organism, that you will begin to distinguish between the kinds of questions you can approach as a scientist and those you cannot, and that this experience will convey the potency and limitations of science. Whether or not you choose a career in a natural science, this course will give you a sense of the excitement of dealing with science as it helps lay the groundwork for an intelligently critical attitude toward science in general and biological science in particular. We urge you to enjoy your laboratory sessions and to make the most of the weeks ahead.

BIOLOGICAL INVESTIGATIONS IN THE LABORATORY

TOPIC 1 MICROSCOPY

Advances in science have often been brought about by the development of instruments capable of extending the normal range of human senses. The light microscope is an example of one of these instruments, and its development brought a major revolution in biological thought by opening up the unexplored world of the animal and plant cell.

The light microscope contains a sophisticated arrangement of magnifying lenses through which light rays, passing from a specimen, are bent and rejoined to produce an enlarged image. In this exercise, you will be using two light microscopes—a compound microscope and a dissecting microscope. You should become familiar with the basic parts and operation of each of these microscopes. You will also be introduced to a variety of living materials and to the proper techniques for studying these materials.

Learning any new technique takes practice and patience. The same is true when learning to use a microscope properly. If your microscope field is not clear and you are straining to see an image, or if you are confused about the specimen you are examining, stop at once to find out what the problem is. If necessary, seek help to correct it.

THE COMPOUND MICROSCOPE

Structure and Operation The compound microscope has two sets of magnifying lenses: the objective lenses located near the specimen, and the ocular or eyepiece lenses at the top of the microscope. This is a delicate and expensive instrument, and must be treated gently. Familiarize yourself with the parts and operation of your instrument, referring when necessary to Figure 1-1.

1. Remove your assigned compound microscope from its cabinet by the **arm**. Holding it upright and supporting the **base** with your other hand, take the microscope to your desk and put it down gently, its arm toward you. A microscope should always be carried in this manner, with *both hands.*

2. The arm supports a vertical **body tube**, which in turn supports the magnifying elements of the microscope. At the top of the tube is the **ocular** or **eyepiece**, one of the lens systems. This part of the microscope is usually removable. The other magnifying elements, called **objectives** (Figure 1-2), are

Note: Page or chapter references for the material covered in each laboratory will be given at the beginning of each topic. They refer to William T. Keeton and James L. Gould, *Biological Science,* 4th ed. (New York: Norton, 1986), and William T. Keeton and Carol Hardy McFadden, *Elements of Biological Science,* 3rd ed. (New York: Norton, 1983).

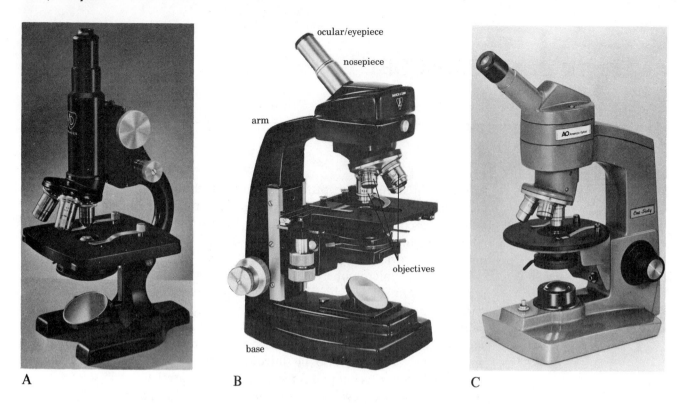

Fig. 1-1. Compound microscopes. *(A, C: Courtesy Reichert-Jung, Inc., a Cambridge Instruments Company; B: Courtesy Bausch & Lomb.)*

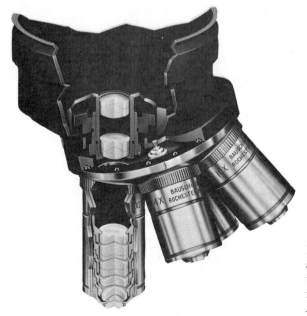

Fig. 1-2. Cutaway of nosepiece assembly and objective, showing the arrangement of optical elements. *(Courtesy Bausch & Lomb.)*

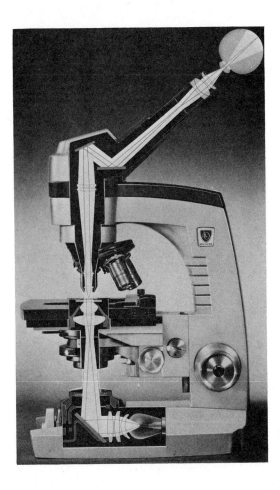

Fig. 1-3. Path of light through a compound microscope. *(Courtesy AO Instrument Company, Buffalo, New York.)*

screwed into a revolving **nosepiece** at the bottom of the body tube. Together, the ocular and an objective constitute the magnifying system of your microscope.

3. Underneath the body tube and nosepiece is a flat plate, the **stage,** upon which the specimen is placed. Directly beneath the stage opening you will notice a system of lenses, the **condenser** (it may be absent in some microscopes), which serves to concentrate light reflected up through it from an attached light or mirror below. On instruments which use a mirror to focus light rays, one side of the mirror is concave and the other is flat. If you do not have a condenser, use the concave surface of the mirror to direct the light rays.

4. Light plays an extremely important role in the operation of a compound microscope. As you can see in Figure 1-3, light is reflected up through the opening in the stage, and passes through the specimen and then into the body tube, ultimately forming an **image** on the retina, the light-sensitive portion of the eye. The critical importance of light makes necessary its careful adjustment, and several controls are available for this purpose. The **iris diaphragm** is attached below the condenser or, if the condenser is absent, just below the stage. Find the **iris-diaphragm control,** a small handle at one side of the diaphragm, and, while looking through the microscope, slide the handle back and forth, noting how the light intensity changes. This control is one of the most important on your microscope. (*Note:* In place of

a diaphragm, some microscopes may have a revolving plate with holes of different sizes to admit varying amounts of light.)

5. Just as with a magnifying glass, viewing a specimen with the microscope requires the lens to be a particular distance from the object. This distance is a property of the lens system and is constant for any particular objective; this distance from specimen to objective is called the **working distance**. At the working distance from an object, an object is in focus. Changes and adjustments in the focus, which are necessary when an object is first placed on the microscope and essential in order to perceive depth, are accomplished by means of **coarse** and **fine adjustment** knobs, usually located on the arm. Try turning each of the knobs, noting carefully how each affects the position of the objectives. The coarse adjustment is used to obtain an approximate focus and the fine adjustment to obtain an exact and a clear focus. Memorize which direction of turning increases the separation between the stage and the objective; initial focusing is always carried out in this direction to prevent damage to the specimen and objective.

Magnification   The magnification of most objectives and oculars is engraved on them. On the ocular, the marking can be found on the top edge or on the smooth cylinder that fits inside the body tube; on the objectives, magnification is engraved on the side of the cylinder. Therefore, if an objective is marked "10×," it magnifies the image ten times in each dimension. The 10× ocular then magnifies this "primary image" another ten times. The image that is finally perceived by the eye has been magnified 100 times the size of the object. The **total magnification** for any combination of objective and ocular is simply the product of the magnifications of each element.

Prepare a table listing the total magnification for each of the combinations of oculars and objectives for the microscope you are using.

Resolution   Resolution is the ability to distinguish two objects or lines as two separate units. Unaided, the human eye has a resolving power of about 100 microns. Therefore, two lines separated by less than 100 microns will appear as a single line. Since the typical animal cell is approximately 20 microns in diameter, a lens system capable of greater resolution than the unaided eye is necessary to view a single cell and the finer details of that specimen. The best compound microscope has a **limit of resolution** of 0.2 microns. The limit of resolution depends on the properties of light. Light of a shorter wavelength such as blue light (400 nanometers) will result in better resolving power than red light (600 nanometers). The resolving power of a lens is described by a number called **numerical aperture**, and is often engraved on the objective. The numerical aperture is calculated from certain physical properties of the lens and the angles at which light enters and leaves it.

In addition to using the terms "low" and "high" as common names of the most frequently used objectives, the objectives may be designated by magnifying power, working distance, or numerical aperture. Compare the values listed in the table below with those on the objectives of your microscope.

COMMON NAME	MAGNIFICATION	WORKING DISTANCE	NUMERICAL APERTURE
Scanning lens	2.5–4	25–55 mm	About 0.10
Low power	10	5–10 mm	About 0.25
High power	40–50	0.15–0.60 mm	0.65–1.00
Oil immersion	90–100	0.05–0.15 mm	About 1.25

USING THE COMPOUND MICROSCOPE

Cleaning the Microscope   Dirt, especially eyelash oil and dust, is apt to accumulate on the surface of your ocular. Since many students use your instrument, make it a habit to clean the ocular each time you use the instrument. Use *only* lens paper made for the purpose, *never* any other material. It is a good practice to clean the objectives as well, since soil, even a small amount that may not be visible to you, seriously interferes with image formation by a compound microscope. The mirror, slide, and cover slip must also be completely clean, otherwise you may get an image that is unclear.

Adjusting the Light   Place the microscope on a firm base, its arm toward you. Fully open the iris diaphragm or its equivalent. Place a 3 × 5–inch index card or piece of white paper over the stage opening. If your light source is not attached to your microscope, position the light about 6 inches away from the mirror and adjust the mirror until you see a bright spot of light on the paper. When you see that spot, the illuminator and mirror are properly adjusted to give the brightest possible illumination, and it will be unnecessary to make any further change in the light except to adjust the intensity with the iris diaphragm.

Focusing   On an index card or a piece of paper, write a lower-case letter "e" about the size of a type character in this book. Put the paper on the stage so that the letter lies right in the middle of the stage opening. Now turn the nosepiece at the bottom of the body tube until the *lowest* power objective clicks into position. Then, *watching from the side of the microscope*, lower the objective until it is about one-quarter inch from the paper. Look into the ocular, and focus slowly upward. As you turn the coarse adjustment, the objective moves upward; when it reaches the correct distance from the paper (What is this distance called?), the letter will be in focus. (*Note:* If you failed to get the letter into focus, it may not have been centered in the stage opening, or you may have "overshot" the working distance by focusing upward too rapidly. Try again, making certain that you *never* move the objective back down toward the slide *without first* looking from the side of the instrument.)

You can avoid a great deal of eyestrain by keeping both eyes *open* while you are looking through the microscope. You will soon learn to disregard the image from the "off" eye by concentrating attention on the microscope field.

1. What is the appearance of the "e" under low power? Is it right side up or upside down?

2. When you move the paper to the right while looking into the microscope, which way does the image appear to move? Which way does the image move when you move the paper toward you? Away from you?

These phenomena are called **inversions**; because of the microscope optics, the object and its movements are inverted. Practice moving the paper or writing and observing different letters until you are used to the inversions.

Changing Magnification   When you want to see more detail, you must switch to a higher power objective. Before doing so, center in the middle of the lower power field the detail you wish to examine. Rotate the nosepiece until the next higher objective clicks into place. Then use the *fine focus* to make the object clear. Unless your specimen is extremely thick or your microscope not equipped with *parfocal* objectives (see note below), it is usually unnecessary to make more than a minor adjustment in the fine focus or to raise the body tube before you change objectives. Make certain, however, that the nose of the higher-powered objective does not touch the specimen. (*Note:* Most microscopes have their objectives machined so that when all objectives are firmly screwed into place, it is unnecessary to make a major change in the focus when magnification is changed. Objectives so aligned are spoken of as being **parfocal** to one another. If your instrument was in sharp focus under the lower power, it should not be necessary to make more than a small change in the fine adjustment to obtain a clear focus after switching objectives.)

If you "lost" the object when you changed magnification, your objectives may not be parfocal or you may not have properly centered the object intended for examination. If the objectives are not parfocal, simply refocus whenever you change objectives. If you did not center properly, try moving the object slightly or change back to the lower power and then recenter.

1. How much area of the "e" can you see under the higher power relative to what you saw under the lower power?

2. How did the brightness of the light change when you changed objectives? If you did not see any change, go back to the lower power and make the change again. If it were necessary to increase the light intensity after switching objectives (as it often is), by what means would you do so?

Field Size   Estimate the size of the visual field of your compound microscope by examining a section of graph paper which has been mounted on a slide.

SCANNING LENS:

LOW POWER:

HIGH POWER:

Depth of Focus   Obtain a slide with three or four colored threads mounted together. Under low power, find a point where several of the

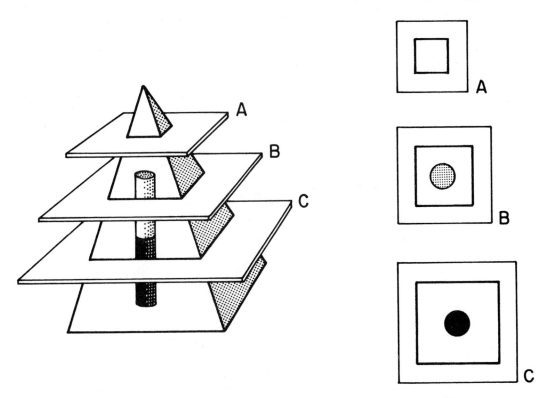

Fig. 1-4. Depth of focus. Sections A, B, and C illustrate how sections taken at different levels through a specimen reveal different structures.

threads can be seen together in the same field. Then, with fine adjustment, slowly focus up and down. Notice that as you do so, different parts of the threads, and different threads, become distinct; when one is in focus, the others, above and below, are blurred. By continuous fine focusing, up and down through the specimen, you can perceive the depth dimension that is not evident when the focus is resting at one point. By focusing up and down, determine the order of the threads, top to bottom, on your slide.

TOP THREAD: _____

_____

_____

BOTTOM THREAD: _____

Examine the threads under high power, and notice that you can see considerably less depth than under low power. Indeed, you may not be able to distinguish one whole thread clearly under high power. As you increase magnification, you can always expect the perceivable depth of the specimen to decrease.

The vertical distance that will remain in focus at any particular point is called the **depth of focus** or depth of field; it is a constant value for each of the objectives. By focusing up and down through specimens that are thicker than the depth of focus, you can get around this limitation in depth perception. Changing the focus in this manner is a little like making successive thin slices of the object and then viewing them one by one (Figure 1-4). Indeed, the structure of some small objects is actually studied by making and observing serially a large number of such slices. Expert microscopists keep themselves continuously aware of depth by making constant changes in the fine adjustment.

SPECIMEN PREPARATION

Many of the specimens you examine will be mounted in water on glass slides. Such a **wet mount** is prepared by placing a drop of material to be examined on a slide or, if the material is dry, by placing it directly on the slide and adding a drop of water. The specimen is then covered with a **cover slip**, a thin piece of glass or plastic that helps prevent the drop from drying and flattens the preparation to avoid the uneven refraction that would occur if light passed through the top of a drop of fluid.

Human epidermal cells

Gently scrape the inside of your cheek with a clean, flat toothpick, and mount the scrapings on a clean, dry slide. Let the scrapings dry, and then add a drop of tap water. Holding a cover slip (with fingers or forceps) by its edges to avoid fingerprinting it, place one side of the cover slip at one edge of the drop. Now gently lower the cover slip over the rest of the drop; if this is done slowly, you will find that water flows evenly beneath the glass. With a little practice, you will learn that the amount of fluid required to mount a specimen depends upon the size and thickness of the specimen: if too much water is added, the cover slip may "float" about on the fluid; and if too little water is used, part of the area under the cover slip may remain dry. You can add more fluid with a dropper at the edge of the cover slip, and withdraw excess fluid by soaking it up with a bit of filter paper or toweling.

Examine the slide under low power, remembering to adjust the light intensity as you view the preparation.

You have removed some of the **epidermal cells** that form a protective surface inside the mouth. Like similar cells all over the body, these are constantly being sloughed off and replaced. Epidermal cells usually come off the cheek in masses, so look for groups of cells; when you have found one, study its margins under high power to see the individual cells. Locate the **nucleus**, a small spherical body in the center of the cell, and the **cytoplasm** (a term that refers collectively to all of the cell's contents apart from the nucleus). The **plasma membrane** bounding the cell is too thin to see with light microscopy (refer to the electron micrographs at the end of this Topic), although you will see the boundary where the cytoplasm ends and the mounting fluid begins. The structure of the cytoplasm is an extremely complicated arrangement of membranes and organelles, but very little of this fine structure can be seen with a compound microscope.

Elodea leaf

With forceps, remove a single leaf from an actively growing *Elodea* shoot. Place the leaf in a drop of water on a microscope slide, and cover the slide with a cover slip. Position the slide so that an area near the midrib is in the center of the field. Change to high power, and adjust the light intensity. With fine adjustment, focus slowly through the layers of cells and observe the following features:

1. **Cell wall**, a rigid, transparent structure that lies outside the cell membrane and is composed mostly of the polysaccharide **cellulose**. Cell walls give plant cells a rigidity absent in animal cells.

2. **Cell vacuole**, appearing as an empty, clear area but actually containing fluid. The vacuole is surrounded by a membrane called the **tonoplast**.

3. **Chloroplasts**, organelles that are suspended in the cytoplasm of the

cell, contain chlorophyll, and serve as the site of photosynthesis. (The nucleus will be difficult to observe in these cells because of the large number of chloroplasts.)

Describe the movement of the chloroplasts in the cells. Determine the number of cell layers in the leaf. Measure and record the size of the cells in each cell layer. Read Appendix 1 on drawing, and make a large drawing of the *Elodea* cell and label the structures that you observe.

Closterium Examination of the unicellular alga *Closterium* will serve as a test of your ability to use the compound microscope properly.

Prepare a wet mount of living *Closterium* cells from the culture provided. Examine the algal cells under all three magnifications, and determine their length and width. Make a drawing of this cell, and identify the **cell wall, nucleus, chloroplast,** and **vacuoles.** Observe the crystals that are in the vacuoles, and describe their movement.

Staining It is often helpful to increase the contrast of certain cell structures, and a great variety of stains is available for this purpose. Some color all parts of cells more or less indiscriminately, while others act more specifically on chemical compounds unique to particular structures within the cell.

Obtain some more cheek cells, place them directly on a clean, dry slide, and allow the scrapings to dry. Place the slide on a paper towel at your desk, and add a drop or two of methylene blue stain. Stains of this type usually kill the cells, but there are a few **vital stains** that may be used with living materials. After two minutes, rinse off the stain with water; this is best accomplished by dripping a little water from your fingers or a dropper over the slide while you hold it above the sink. Dry the bottom of the slide, add a drop of water and a cover slip, and observe carefully.

What features of the cheek cell stain most prominently? Estimate the size of the cheek cells.

USING OIL-IMMERSION OBJECTIVES

The working distance of an oil-immersion (90–100X) objective is less than 1 mm, which means that the end of the objective must almost touch the slide for the specimen to be in focus. Since the oil-immersion field is much smaller than the high-power field and the depth of focus with oil immersion is also very small, the light intensity almost always needs to be increased considerably.

In order to obtain maximum resolution from an oil-immersion objective, the layer of air between the objective and the specimen is replaced by a drop of oil into which the objective is carefully immersed. This procedure eliminates much of the refraction and loss of light that occurs when light passes from a slide into the air before reaching the lens of a dry objective.

If the oil-immersion objective is not already attached to your microscope, obtain one from the instructor. Remove the objective carefully from its case, and put the case in a safe place. Screw the objective into the nosepiece of your microscope.

Examining bacterial cells

Using the slide of human epidermal cells that you prepared, focus on the tiny specks between these cells. These are the bacterial cells which are normally found in the mouth. After you have located the bacterial cell under high power, center it carefully in the high-power field and then follow the directions below.

1. Raise the body tube enough above the slide so that you can easily swing the oil-immersion objective into position. Deposit a drop or two of immersion oil on the slide as well as on the condenser lens, if you have a condenser. Click the objective into position.

2. While watching from the side, slowly lower the objective until its tip makes contact with the oil. Then, looking very closely at the slide, lower the objective a little more, until you can see that it is nearly touching the slide.

3. While looking into the ocular (with both hands on the fine adjustment), very slowly raise the objective. As you do this, you should bring the object into focus. If you do not, you may not have gotten the objective close enough to the slide; try again, asking for help if you continue to have difficulty.

Diagram the various types of bacterial cells that you observe on your slide. How many different cell types did you find?

After you have finished making your observations, raise the body tube and wipe off the slide and objective with lens paper to remove most of the oil. The remaining oil can be cleaned off with a little isopropyl alcohol.

THE DISSECTING MICROSCOPE

Structure and Operation   The other microscope you will be using in this course is the binocular dissecting microscope. As the term "binocular" implies, this microscope has a pair of oculars, one for each eye. Most dissecting microscopes also have a double objective system, arranged in such a way as to render a stereoscopic (three-dimensional) image. Figure 1-5 illustrates two commonly used dissecting microscopes.

Notice that the distance between the oculars can be adjusted. Pull them apart or push them together until you have them adjusted to the correct position for your eyes. What is the magnification of the oculars? (If your microscope has an ocular on which the focus can be adjusted, look through the *nonadjustable* ocular first—with your other eye closed—and obtain a clear focus on some object in the usual manner. Then open the other eye, close the first, look through the adjustable ocular, and rotate the milled cuff around it until you obtain a clear image.)

Since the magnification of the dissecting microscope is lower than that of your compound microscope, the working distance is greater; hence much larger objects can be examined. Some dissecting microscopes have objectives of several different magnifications mounted on a revolving turret. If your instrument is of this type, determine the power of each objective. Other microscopes may have a rotating dial in place of a turret or an adjustment knob atop the body; when turned, it permits continuous change in the magnification throughout a range marked on the knob. This latter type is called a "zoom" microscope. Whatever the arrangement on your instrument, determine how the magnification may be changed. What is the lowest possible magnification? The highest?

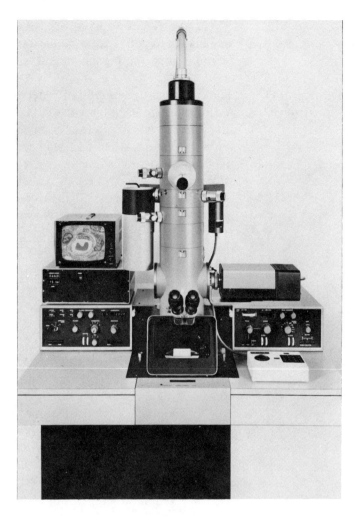

Fig. **1**-5. Transmission electron microscope. *(Courtesy Carl Zeiss, Inc., Thornwood, New York.)*

***Practice with the
dissecting microscope***

1. With forceps, obtain a small moss plant from the clump at the supply desk, and place the plant on a dry slide without using a cover slip.

If your microscope is equipped with a substage mirror and a transparent stage, adjust the illuminator to see whether you get a clearer view of the moss with or without illumination from beneath. Light passing through an object and then into the microscope is **transmitted** light, while light going from the illuminator directly to the object (and reflected by the object up into the microscope) is **reflected** light. One of the decisions a microscopist must make is whether to use transmitted or reflected light. Which form have you already used with the compound microscope?

Whether you use reflected or transmitted light depends on the size and opacity of your specimen. If you decide to use reflected light, remember that the illumination must be more intense, since more of the light is lost by scattering before it reaches the microscope lenses.

If the stage of your dissecting microscope has a black side and a white side, determine which is better for examination of the moss.

Examine the moss carefully, turning it several times so as to change its position. How would you describe the gross structure of a moss plant?

2. Net a guppy from the aquarium in the laboratory, and place it in a small covered petri dish of aquarium water. Examine the structure, gill movements, and swimming movements of the fish under the dissecting microscope.

Choosing the proper instrument

The usefulness of microscopy depends a great deal upon the wisdom, imagination, and good judgment of the person using the instrument. The person must decide when to use a compound microscope, when to use a dissecting microscope, and when to use other types of magnification (if available). The decision must be made as to what magnification to use when the magnification is adjustable.

Look at the moss with the compound microscope (without cover slip) and with a hand lens, if one is available. Which of these instruments do you find the most useful for an object like the moss, which is opaque, three-dimensional, and large enough to be seen with the naked eye? Which instrument would you use to determine the cellular structure of a single moss leaf?

Perform the following, deciding in each case which microscope is most appropriate.

1. With a razor blade, make very thin sections—transparent, if possible—of potato tuber, and examine the cells that compose the tuber. The oval bodies in these cells are starch storage organelles, called **leucoplasts.** Stain a section with a drop of iodine (placed at the edge of the cover slip), and note the resulting color in the leucoplasts. This color is a positive test for starch.

2. Examine a drop of blood, obtained from your finger with a sterile lancet and diluted on a slide with two drops of 0.9% salt solution.

3. Make cross and longitudinal sections of the plant stems provided by your instructor, and examine them for general appearance and cellular structure.

4. Examine the organisms in pond water provided by your instructor.

OTHER TYPES OF MICROSCOPY

One can arrange a light microscope for use with polarized light, dark field, fluorescence, phase contrast, and a number of other types of microscopy that render some cell components very clearly. Some of these may be on demonstration, with explanatory material beside them.

As you already know, a great deal has been discovered about the internal "fine" structure of cells. The limitations of light as a refractable medium and the practical problems of manufacturing very tiny lenses make it impossible to attain magnifications higher than about 2000X with the light compound microscope. Most of our knowledge of cellular fine structure has, therefore, been derived from **electron microscopy,** a technique similar in principle to light microscopy, but one that employs beams of electrons instead of light and magnets in place of glass lenses. Electron beams have a much shorter wavelength than visible light, and refracting them with magnets can yield resolutions several hundred to several thousand times higher than conventional light microscopy.

An electron microscope is illustrated in Figure 1-5, and Figure 1-6 shows an electron micrograph of a common type of cell surrounded by smaller micrographs of the various common organelles. Carefully label the cell parts in Figure 1-6.

Examine other microscope equipment on demonstration.

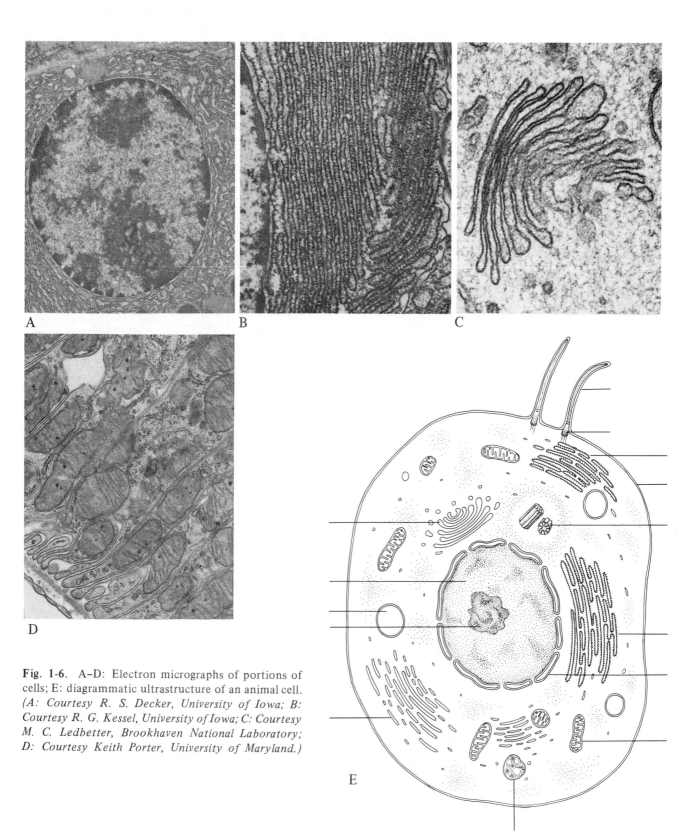

Fig. 1-6. A–D: Electron micrographs of portions of cells; E: diagrammatic ultrastructure of an animal cell. *(A: Courtesy R. S. Decker, University of Iowa; B: Courtesy R. G. Kessel, University of Iowa; C: Courtesy M. C. Ledbetter, Brookhaven National Laboratory; D: Courtesy Keith Porter, University of Maryland.)*

QUESTIONS FOR FURTHER THOUGHT

1. What is meant by the term "parfocal"?

2. What is meant by "depth of focus"? With which objective of your compound microscope would depth of focus be greatest?

3. What means, other than the iris diaphragm, could you employ to control the light entering a compound microscope?

4. By what means, besides changing objectives, could you alter the total magnification of a microscope?

5. Why is it necessary to center a detail in the low-power field before switching to a higher power?

6. Is the cell wall of a plant cell alive? Where does it come from?

7. Define or describe and explain "stain," "primary image," "ocular."

TOPIC 2 PROCARYOTIC AND EUCARYOTIC CELLS

Biological Science, 4th ed.
Chap. 5, pp. 111–136, 249–251, 550,
1013–1017, 1038–1040, 1047
Elements of Biological Science, 3rd ed.
Chap. 4, pp. 81–96, 157–159, 611–613,
628, 633

All living organisms are composed of cells, complex units with highly specialized internal structures that fulfill a variety of important functions. With the development of the electron microscope and improved techniques for specimen preparation, biologists have been able to distinguish two different types of cells: procaryotic cells and eucaryotic cells. Bacteria and Cyanobacteria, also known as blue-green algae, are composed of procaryotic cells, while eucaryotic cells are found in protists, fungi, animals, and plants. Procaryotic cells, which are far less complex than eucaryotic cells, are generally smaller in size, contain few or no cytoplasmic organelles, and divide by binary fission. Both cell types share a number of characteristics in common, which implies a common evolutionary origin. Procaryotic and eucaryotic cells both use similar types of enzymes, transmit genetic information via DNA, and are surrounded by a plasma membrane which regulates the exchange of materials in and out of the cells.

We will examine the structure of some procaryotic cells, a number of eucaryotic cells and cell structures, and an organism that has acquired Cyanobacteria as symbionts.

PROCARYOTES

Cyanobacteria   The Cyanobacteria exhibit primitive structure and reproduction; they illustrate a type of cellular organization from which certain more complex cells—plant cells—probably evolved. These organisms are photosynthetic like algae and higher plants. Unlike these groups, however, Cyanobacteria cannot carry on respiration and photosynthesis simultaneously; respiration takes place in Cyanobacteria only during the dark.

Examining Oscillatoria

This filamentous Cyanobacterium is commonly found on stones or on the surface of damp earth.

1. Obtain some *Oscillatoria,* and make a wet mount for microscopic examination. (If the culture is on agar, be sure not to remove any agar when you take the specimens.) New cells are added to the filament through binary fission; sexual reproduction has never been observed in the Cyanobacteria. Vegetative reproduction occurs when filaments break apart, often following the formation of a gelatinous region between a pair of cells. Such regions may often be seen under high power. Do cells of *Oscillatoria* divide in more than one plane?

2. Look closely at a single cell, and notice the absence of nucleus and chloroplasts. Like bacteria, these organisms are termed procaryotic because

15

neither the genetic material nor the photosynthetic pigments are located within membrane-enclosed organelles in the cytoplasm. Rather, the photosynthetic pigments of *Oscillatoria* are located in flattened membranous vesicles called thylakoids, not visible with the light microscope.

The cells in *Oscillatoria* are all alike, and the filament may indeed be regarded as a colony of individual plants. What criteria would you use to distinguish colonies from true multicellular organisms?

Other Cyanobacteria, like *Gloethece,* are unicellular and bound together in prominent gelatinous sheaths. What advantages might there be in this type of organization (in contrast to individual free-living cells)?

3. Find an area in which several *Oscillatoria* filaments cross, and observe the peculiar gliding motion. The mechanism of this movement is unknown. Many other algae show movements of the individual cells, colonies, or cell organelles (such as chloroplasts, which, in some algae, orient to the changing angle of the incident sunlight). How might such movements be adaptive?

Examining Gloeocapsa

This Cyanobacterium is normally found in thick mats on damp rocks. Examine living specimens. What is the appearance of the individual cells? How does the cellular structure compare to that of *Oscillatoria*? Are there isolated single cells or only colonies? How many cells occur in the colonies, and what holds them together? Is the binding material apparent around single cells or only around the colony? Sketch several specimens of *Gloeocapsa.*

PROCARYOTE TO EUCARYOTE

Some biologists believe that eucaryotic cells evolved from procaryotic cells through a gradual and continuous process of mutation and natural selection. Others believe that modern eucaryotic cells evolved from a now-extinct ancestor, a urcaryote, which itself arose when certain cells acquired Cyanobacteria and aerobic bacteria as symbionts. This latter hypothesis, which assumes several important changes to the genetic material of the organisms involved, is called the endosymbiotic hypothesis. As we shall see, the symbiotic association of Cyanobacteria with eucaryotic cells is not an unusual phenomenon.

Procaryote and eucaryote symbiosis

1. Using a clean slide and cover slip, prepare a wet mount of *Glaucocystis nostochinearum.* This organism is the result of a symbiotic relationship between an algal cell, a eucaryote, and the Cyanobacterium *Skujapelta nuda.*

2. Observe your preparation with the compound microscope at low and high power. Make detailed drawings, and label cell parts.

What would be the selective advantage of this relationship to each organism?

EUCARYOTIC CELLS

Examining Amoeba

Amoebae are among the simplest of the protists. They are abundant in fresh and salt water, and in moist soil. An *Amoeba* possesses pseudopodia, or false feet, which are cytoplasmic extensions used for locomotion and for engulfing food particles.

1. Take a clean slide to your instructor, and obtain a drop or two of culture with several living *Amoebae*. First observe your preparation without a cover slip. Locate the organisms, using low power of the compound microscope. You may have to decrease the light intensity. The *Amoebae* will look like grains of sand in the viewing field of your microscope. Note the general size of the organisms, changes in shape, and method of locomotion. Locomotion of this type, common to many kinds of cells, including some of your white blood cells, is called **amoeboid movement.**

2. Now add a cover slip. You should support the cover slip either with small pieces of broken cover slip or with Vaseline. The latter is preferable since it serves the double function of raising the cover slip (so that the specimens are not crushed) and protecting the preparation from drying out (Figure 2-1).

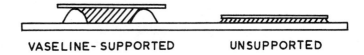

VASELINE- SUPPORTED   **UNSUPPORTED**

Fig. 2-1. Vaseline-supported and unsupported cover slips. Vaseline is a good means of supporting the cover slip when a relatively thick or fragile specimen is to be studied. It may be applied either to the slide or to the cover slip with a toothpick or with a syringe into which the jelly has been drawn after melting.

3. Under low and high powers, identify:
   a. Cytoplasm—containing many granules.
   b. Nucleus—a slightly ovoid structure near the center of the cell.
   c. Food vacuoles—small, dark, food-filled areas in the cytoplasm.
   d. Contractile vacuole—a clear, spherical area, filled with fluid and often near the part of the organism farthest from that actually moving forward in a particular direction.

Remember that the plasma membrane is too thin to see, although you will be able to see the boundary between the cytoplasm and the mounting medium.

It may be necessary to decrease the light intensity before you can see these structures clearly, especially under low power. High power will show more detail. Watch especially the contractile vacuole, which will be seen to pulsate at regular intervals if it is closely observed. What is the function of this behavior of the contractile vacuole?

Watch carefully as the *Amoeba* moves along the glass surface. You will note that the shape of the cell can change constantly and that the granular cytoplasm inside does not remain still, but is commonly seen in a streaming motion. Observe the streaming under high power, as the animal flows along with outward bulgings of the cell called "pseudopodia" (singular, "pseudopod"). In the flow of cytoplasm associated with the formation of a pseudopod, you should be able to distinguish:
   e. A clear outer layer of cytoplasm.

f. Two inner granular layers of cytoplasm (the outer composed of a stiffer, gel-like nonmoving ectoplasm, and the inner endoplasm in motion).

If you have any difficulty seeing these layers, check your slide preparation and the microscope optics for cleanliness; if you continue to have trouble, ask your instructor for help. Study carefully the formation of a pseudopod, particularly the transition from endoplasm (granules in motion) to ectoplasm (granules fixed in position). At what point in the formation of a new pseudopod does the transition take place? Make sketches, if you like, and continue these observations for as long as you wish. (*Note:* If you watch carefully, you may see an *Amoeba* ingesting one of the small, ciliate protozoans upon which it feeds. The food is engulfed or surrounded by advancing pseudopodia, and a vacuole is thus formed around it. Digestive enzymes are then secreted into the vacuole. If you can find a newly formed food vacuole—the prey will still be moving inside—watch it for a while to determine how long it takes before the prey is inactivated.)

If a film of amoeboid movement is available, make sure that you see it sometime during the period.

Before going on to the next section, use your text to help you identify the phylum and class of *Amoeba,* and enter it in the appropriate portion of Appendix 2.

***Examining* Nitella**

Another example of an organism with eucaryotic cells is the green alga *Nitella.* This organism inhabits fresh or brackish waters and has giant cylindrical cells that may range up to 5 cm in length. *Nitella* cells may contain one or more nuclei.

1. Obtain a strand of *Nitella* from the tray at the supply desk. Examine the specimen with the dissecting microscope. The long pieces of "stalk" that separate tufts of lateral "branches" are called "internodes," and in *Nitella* the internodes are actually single cells. A large balloonlike **vacuole** in each of the cells makes it appear transparent. Bounding the *Nitella* cell, as in nearly all plant cells, there is a firm **cell wall.**

2. Place the petri dish on your compound-microscope stage, being careful not to get water on the stage or objectives, and focus carefully with low power on one of the internodes. Observe the **chloroplasts,** structures that carry out photosynthesis, which lie side by side and are arranged in perfect rows along the inner surface of the cell wall. Contrary to appearances, the chloroplasts actually reside in the cytoplasm; the cytoplasm is pressed firmly against the cell wall by the vacuole, which occupies most of the volume of the cell. One peculiar feature is the conspicuous absence of chloroplasts in the pair of so-called neutral lines that run in a loose spiral around the internode.

The chloroplasts are stationary in the cytoplasm, but if you focus properly just beneath this layer, you can see cytoplasmic streaming. If you look very closely, you will see that the cytoplasm is streaming in opposite directions on either side of a neutral line. At the end of the internode, in favorable preparations, the streaming particles can be observed to glide around from one side of the neutral line to the other. What functions do you think such streaming may serve? In older cells, the nucleus will have divided into several portions that travel about in the streaming cytoplasm. Studies on organisms such as *Amoeba* and *Nitella* suggest that actin filaments play a role in cytoplasmic streaming.

UNKNOWNS

Your instructor will make available to you cultures of several procaryotes and eucaryotes.

1. Prepare a wet mount of each culture, and examine each carefully with the compound microscope under low and high power. Remember to adjust the light intensity on your microscope as you make your observations.

2. Make a sketch of each organism. Label cell parts, determine the size of the cells, and decide whether the organism is composed of procaryotic or eucaryotic cells. Justify your answer.

TOPIC 3 ENZYME ACTIVITY

A cell is the site of constant chemical activity as energy is extracted (as ATP) from the catabolism of large organic molecules and put to work in numerous energy-requiring activities. The mere maintenance of its own complexly ordered structure is a chemical task of staggering magnitude, but in higher organisms a cell usually has some specific activity, such as hormone or special enzyme secretion, active transport, etc. If it is a green-plant cell, it also carries out the complex biochemical synthesis of carbohydrates, from carbon dioxide and water. All the cell's activities, in fact, are based upon chemical reactions, and all these reactions are made possible by enzymes.

Enzymes are large globular proteins that act as catalysts in biochemical reactions. A catalyst, you recall, affects reaction *rate* by reducing the activation energy necessary to initiate a reaction (Figure 3.35, *Biological Science*) but cannot change the direction of the reaction or the final equilibrium. Enzymes have a **specificity** due to the unique three-dimensional configuration created by the specific folding of the particular amino-acid chain. Only a few of the amino acids are involved in the reactive portion of the enzyme, which is called the **active site**. Because of its specificity, an enzyme tends to accelerate changes in only one reactant or a group of related reactants, called **substrates**. The result is that many different enzymes may be present in the cell, acting simultaneously and without mutual interference.

According to current theories of enzyme activity, the enzyme (E) is pictured as binding temporarily to its substrate (S), the substance upon which the enzyme acts, to form a complex (ES). As a result of this transient union, the energy required to activate the breakdown of the substrate molecule is reduced so that it breaks apart more easily into the products of the reaction (A and B). In summary:

$$E + S \longrightarrow ES \longrightarrow A + B + E$$

Note that the enzyme is recovered at the end of the reaction and is, therefore, available to catalyze the breakdown of additional substrate molecules.

In this exercise you will study the enzyme **catalase**, which accelerates the breakdown of hydrogen peroxide (a common end product of oxidative metabolism) into water and oxygen, according to the summary reaction

$$2H_2O_2 + catalase \longrightarrow 2H_2O + O_2 + catalase$$

This catalase-mediated reaction is extremely important in the cell because it prevents the accumulation of hydrogen peroxide, a strong oxidizing agent which tends to disrupt the delicate balance of cell chemistry.

Catalase is found in animal and plant tissues, and is especially abundant in plant storage organs such as corms and potato tubers, and in the fleshy parts of fruits. You will isolate catalase from potato tubers and measure its rate of activity under different conditions. A glass-fiber filter will be immersed in the enzyme solution, then placed in the hydrogen-peroxide substrate. The oxygen produced from the subsequent reaction will become trapped in the disc and give it buoyancy. The time measured from the moment the disc touches the substrate to the moment it reaches the surface of the solution is a measure of the rate of enzyme activity.

EXPERIMENTS WITH CATALASE

Enzyme activity is controlled by a number of factors, all of which affect the efficiency with which the enzyme binds to the substrate. Enzyme and substrate concentration, as well as changes in temperature, can influence the frequency of enzyme-substrate collisions. The pH or salt concentration affects the amino-acid side chains on the binding site and can alter the overall conformation of the enzyme. Enzyme control can also involve substances which are so similar to the normal substrate molecule that they can bind to the active site, preventing normal enzyme activity. In this lab, you will examine some of these factors and determine their effect on catalase activity.

Extraction of catalase

1. Peel a fresh potato tuber, and cut the tissue into small cubes. Weigh out 50 gm of tissue.
2. Place the tissue, 50 ml of cold distilled water, and a small amount of crushed ice in a prechilled blender.
3. Homogenize for 30 seconds at high speed.

FROM THIS POINT ON, THE ENZYME PREPARATION MUST BE CARRIED OUT IN AN ICE BATH.

4. Filter the potato extract, then pour the filtrate into a 100-ml graduated cylinder. Add cold distilled water to bring up the final volume to 100 ml. This extract will be arbitrarily labeled 100 units of enzyme per ml (100 units/ml) and will be used in the tests.

Effect of catalase concentration

Before considering the factors which affect enzyme reactions, it is important to demonstrate that the enzyme assay verifies that catalase is an enzyme and that it behaves according to the rules of enzyme kinetics. One way to demonstrate this is by determining the effect of enzyme concentration on the rate of activity, while using enough substrate so that the reaction is not limited.

Label four 50-ml beakers as follows: 100, 75, 50, and 0 units/ml. Prepare 40 ml of enzyme for each of the above concentrations in the following manner:

ML ORIGINAL ENZYME	+ ML COLD DISTILLED H_2O	= UNITS/ML
40	0	100
30	10	75
20	20	50
0	40	0

Keep your catalase preparations in the ice bath. Label an identical set of beakers for the substrate. Into each of these beakers, measure out 40 ml of a 1% hydrogen-peroxide solution.

Using forceps, immerse a 2.1-cm glass-fiber filter disc to one-half its diameter in the first of the catalase solutions you have prepared. Allow the disc to absorb the enzyme solution for 5 seconds, remove, and drain for 10 seconds on a paper towel. Drop the disc into the first substrate solution. The disc will sink rapidly into the solution. Then the oxygen produced from the breakdown of the hydrogen peroxide by catalase will become trapped in the fibers of the disc and cause the disc to float to the surface of the solution. The time (t), in seconds, from the moment the disc touches the solution to the moment it again reaches the surface indicates the rate (R) of enzyme activity, where $R = 1/t$. Repeat the procedure twice for each enzyme concentration, and average the results. Record your data in Table 1, and plot your results on Graph A.

How does enzyme activity vary with enzyme concentration?

Effect of substrate concentration

To determine the effect of substrate concentration on enzyme activity, first label eight 50-ml beakers as follows:

0	0.8% H_2O_2
0.1% H_2O_2	1% H_2O_2
0.2% H_2O_2	5% H_2O_2
0.5% H_2O_2	10% H_2O_2

Add 40 ml of the proper (as outlined above) H_2O_2 solution to each beaker. Make sure that the substrate solutions reach room temperature before beginning your assay. Next, use the procedure previously described, in which each filter disc absorbs enzyme extract and is then dropped into a substrate solution. Record your results in Table 2. Repeat the procedure twice for each substrate concentration, and average the results. Plot your results on Graph B, labeling each axis carefully.

How is the rate of enzyme activity affected by increasing the concentration of the substrate? What do you think would happen if you increased the substrate concentration to 20% H_2O_2? Does changing the substrate concentration have the same effect as changing the enzyme concentration?

Effect of enzyme inhibition

Hydroxylamine attaches to the iron atom (a part of the catalase molecule) and thereby interferes with the formation of enzyme-substrate complex. Add 5 drops of 10% hydroxylamine to 1 ml of enzyme extract, and let the solution stand for 1 minute. Prepare a control solution to test. Then measure the activity of each solution. Use 40 ml of 1% H_2O_2 for the substrate. Explain the results.

Effect of temperature

Using 40 ml of a 1% hydrogen-peroxide solution as the substrate, and enzyme extract of 100 units/ml, measure the enzyme activity in the usual manner. Run the reactions in water baths at different temperatures celsius, such as 4°, 15°, room temperature (which is about 22°), 30°, and 37°. The catalase and substrate should be brought to the testing temperature *before* they are used. Record the exact temperature and your data in Table 3, and plot the results on Graph C. Also test the activity of enzyme that has been boiled.

From these data, what can you conclude about how temperature affects enzyme activity? How would you explain the results?

ENZYME CONCENTRATION (UNITS/ML)	TIME TO FLOAT DISC (SECONDS)				RATE
	TRIAL 1	TRIAL 2	Σ	χ	

TABLE 1. DATA FROM EFFECT OF ENZYME CONCENTRATION ON RATE OF ACTIVITY.

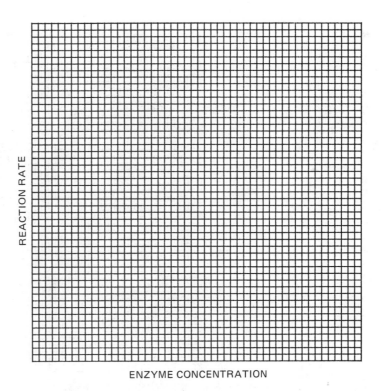

GRAPH A. EFFECT OF ENZYME CONCENTRATION ON RATE OF ACTIVITY.

SUBSTRATE CONCENTRATION	TIME TO FLOAT DISC (SECONDS)				RATE
	TRIAL 1	TRIAL 2	Σ	χ	

TABLE 2.   DATA FROM EFFECT OF SUBSTRATE CONCENTRATION ON CATALASE ACTIVITY.

GRAPH B.   EFFECT OF SUBSTRATE CONCENTRATION ON CATALASE ACTIVITY.

TEMPERATURE	TIME TO FLOAT DISC (SECONDS)				RATE
	TRIAL 1	TRIAL 2	Σ	χ	

TABLE 3. DATA FROM EFFECT OF TEMPERATURE ON CATALASE ACTIVITY

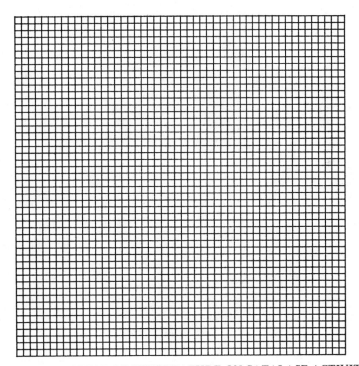

GRAPH C. EFFECT OF TEMPERATURE ON CATALASE ACTIVITY

QUESTIONS FOR FURTHER THOUGHT

1. From your data, what is the optimal temperature for catalase activity?
2. How could you break down hydrogen peroxide without using an enzyme?
3. What reaction occurs between catalase and hydroxylamine?
4. By what mechanism does hydroxylamine interfere with catalase activity?

TOPIC 4 AUTOTROPHIC NUTRITION

Biological Science, 4th ed.
Chap. 8, pp. 140–145, 195–217, 221–237
Elements of Biological Science, 3rd ed.
Chap. 6, pp. 97–102, 109–123, 139–150

Organisms require a constant supply of usable energy. This is a direct consequence of the Second Law of Thermodynamics, which maintains that all systems have a natural tendency toward increasing disorder, a tendency that can be counteracted only by the expenditure of energy. Since any system in a living organism is highly ordered, it follows that energy within this system would soon be depleted were there no external energy source such as the sun. Only certain organisms, however, have the capacity to trap the energy of sunlight directly and to convert it into potential chemical energy; this energy-trapping process is called **photosynthesis**. Although terrestrial plants, the photosynthetic organisms with which you are probably most familiar, are usually green, there are many important photosynthetic organisms that are not green—for example, the brown and red algae (common seaweeds). What is common to all photosynthetic organisms, however, is the green pigment **chlorophyll**, a compound that can be activated by light. Photosynthesis, as you know, is the conversion (often called "fixation") of carbon dioxide and water into organic compounds. The energy that drives this process comes from light-activated chlorophyll.

What sets photosynthetic organisms apart from the rest of the living world is their ability to synthesize from inorganic compounds and from the energy of the sun all of the organic compounds they require for building and maintaining their protoplasm. Organisms having this independent nutritional status are known as **autotrophs** (self-feeders). All other living organisms must depend upon the organic compounds synthesized by autotrophs; these compounds provide them with energy (through respiration and fermentation) and protoplasmic building blocks. Organisms having this dependent nutritional status are called **heterotrophs**. Why autotrophic organisms must be the first link in any food chain and, therefore, assume a central and crucial role in the community of life should now be clear to you. How do you think the organic carbon synthesized by autotrophs becomes available to heterotrophs?

In this unit you will examine leaf structure and make some quantitative measurements of photosynthesis. Although you will mainly study terrestrial plants, you should not forget the importance of marine algae as the primary fixers of carbon. You should also remember that a small number of autotrophs are chemosynthetic rather than photosynthetic; they are able to extract energy from certain inorganic compounds, as, for example, the bacteria found in stagnant ponds that combine hydrogen gas and sulfur to produce the foul-smelling gas hydrogen sulfide.

LEAF STRUCTURE

Ligustrum   The leaf as the principal organ of photosynthesis must be adapted to expose large numbers of chloroplasts to sunlight. The leaf must also provide a moist cell surface for exchange of gases with the environment yet prevent excessive water loss. Finally, it must contain extensions of the vascular system of the stem to move the products of photosynthesis from their sites of production to the nonphotosynthesizing parts of the plant and to move water and minerals into the leaves. As you examine a representative leaf in cross section, ask yourself which of its structural features are of particular importance in making it an efficient site for photosynthesis.

Obtain a prepared slide of a leaf cross section from the plant *Ligustrum* (privet hedge) or another dicot leaf (Figure 4-1). Examine the section under low and high power. Note that the leaf is composed of a complicated arrangement of cells. You will probably first notice sections of vascular bundles, recognizable on the whole leaf as **veins**. Focus on the main "midrib" of the leaf in the center of the section. You will be able to see an exact cross section of a vascular bundle. Since the veins in the leaf run in a complex pat-

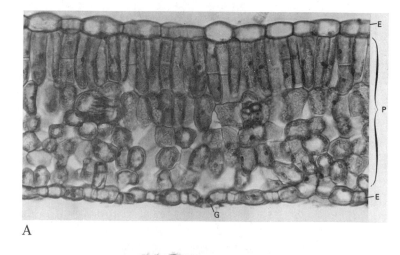

A

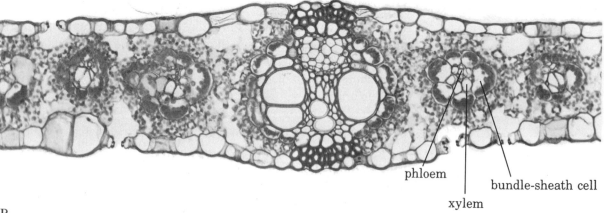

B

Fig. 4-1. Leaf cross sections. A: Photograph of privet leaf (E: epidermis; G: guard cell; P: parenchyma); B: photograph of a corn leaf. *(A: Courtesy Thomas Eisner, Cornell University; B: Courtesy Carolina Biological Supply Company.)*

tern, the section often cuts them obliquely, giving a "smeared appearance." A vascular bundle is composed of two types of vascular tissue: xylem and phloem. These are continuous with the same tissues in the stem and root. Between the veins lie extensive regions of photosynthetic tissue. Observe (from the top of the leaf down):

1. The **cuticle**, a waxy layer on the upper and lower leaf surfaces but, in *Ligustrum*, thicker on the upper surface. What is its function? Why is it thicker on the upper surface?

2. The **upper epidermis.** How many cells thick is the upper epidermis? Does this layer contain chloroplasts?

3. The **mesophyll**, that region between the upper and lower epidermal layers, excluding the vascular tissues. Note the **chloroplasts.** Where within individual cells are they located? Chloroplasts contain several different green or blue-green **chlorophyll** pigments and also yellow or orange pigments called **carotenoids**.

The electron micrograph of a chloroplast (Figure 4-2) reveals that the chloroplast has a complex internal membranous organization; each group of membranes, which looks like a stack of several coins, is called a **granum.** The lower inset shows a granum, and the upper inset diagrams the structure thought to characterize the layers within the granum. You will note that chlorophyll is bound between the layers.

What photosynthetic reactions occur in the chloroplast? Can any of them occur in chloroplasts outside a living plant? Can any of them occur in chloroplasts that are not intact? Would you consider an isolated chloroplast a living unit? Because the mesophyll contains so many chloroplasts, it is the primary photosynthetic tissue of the plant. It has two distinct regions: **palisade mesophyll**, cylindrical cells nearest the upper epidermis; and **spongy mesophyll**, a region of irregularly shaped cells beneath the palisade layer, containing many intercellular spaces.

4. The **lower epidermis.** Do these cells contain chloroplasts? Locate pairs of very small cells called **guard cells**, bordering **stomata** (pores) leading to intercellular spaces in the mesophyll. Are there chloroplasts in the guard cells? How do the guard cells function?

Plants that grow in temperate climates have leaves structually similar to leaves of privet (*Ligustrum*) that you have just examined. These plants are called C_3 plants because of the nature of their biochemical pathway. In these plants, carbon fixation takes place in the mesophyll cells.

Zea mays

Plants that must survive in drier climates and carry on photosynthesis under intense heat and low carbon-dioxide levels have special adaptations. These plants, such as corn, some grasses, and sugarcane, are called C_4 plants because the first product after the addition of CO_2 is a four-carbon compound. Carbon fixation occurs in the bundle-sheath cells that surround the vascular bundles rather than in the mesophyll cells, as in C_3 plants.

Examine a prepared slide of a cross section of a corn (*Zea mays*) leaf (Figure 4-1B). Scan the section under low power, and observe the vascular bundles. Note that all the bundles are visible only in cross section because of the arrangement of veins in the leaf. How does the mesophyll layer differ from that of the *Ligustrum* leaf? In corn, the mesophyll layer contains the enzymes needed to pump in carbon dioxide under low CO_2 tension. Note

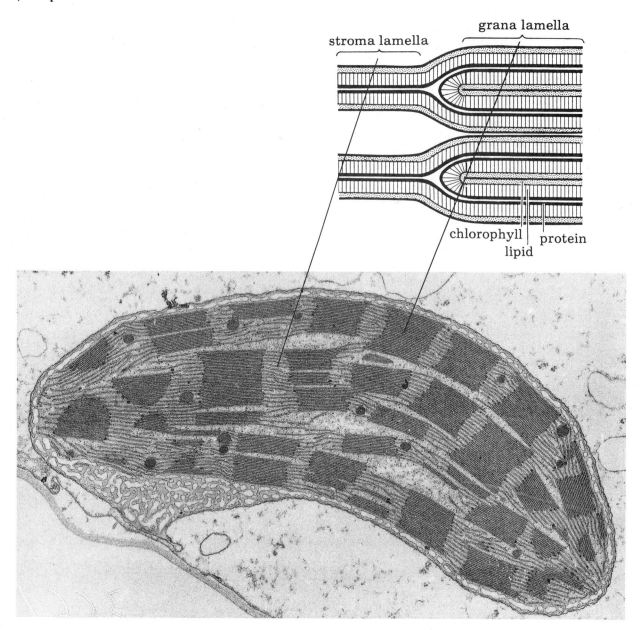

stroma lamella

grana lamella

chlorophyll
lipid
protein

Fig. 4-2. Structure of the chloroplast. The electron micrograph shows the structure of a whole chloroplast from a cell of a corn leaf. The interpretive sketch shows membranes of grana and stroma lamellae. *(Micrograph: Courtesy L. K. Shumway; diagram: Modified from A. J. Hodge et al., Biophysics Biochem. Cytol., vol. 1, 1955.)*

that the bundle-sheath cells in corn contain chloroplasts, while they were absent in the *Ligustrum* bundle-sheath cells. Compare the size of the bundle-sheath cells in privet and corn. Why would the size of these cells be important? What other structural adaptations can you find in the corn section that would make it suitable for survival in drier climates? Note the importance of these adaptations.

Stomatal structure Review the relationship of a stoma to the leaf interior, and the structure of the guard cells by examining the photomicrographs (Figure 4-1A and 4-1B) of cross sections through privet (*Ligustrum*) and corn (*Zea*) leaves.

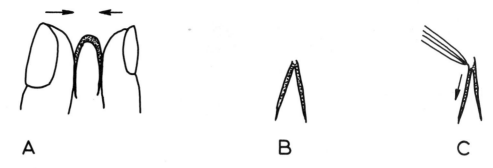

Fig. 4-3. Method for removing epidermis from leaf. A: Pinching the leaf; B: leaf after epidermis cracks; C: removing epidermis with forceps.

Make a wet mount of the lower epidermis from *Rhoeo discolor* or other leaves supplied by your instructor. Figure 4-3 shows the method for removing the epidermis from the leaf. Prepare and label a drawing of guard cells and stoma in *surface* view.

Notice in the photomicrographs that each stoma leads into a maze of chambers in the mesophyll. It is across the moist walls of these mesophyll cells that exchanges of water, carbon dioxide, and oxygen can take place. Why don't the mesophyll cells dry out?

The size of the stoma is regulated by changes in shape of the guard cells: when the guard cells are turgid, the stoma is open; when the guard cells are flaccid, the stoma is closed (Figure 4-4). Guard-cell shape is controlled by water content; this, in turn, is influenced by the concentration of osmotically active particles in the guard cells. The most important solute is sugar, the concentration of which is changed by photosynthesis and by the conversion of starch into sugar or sugar into starch (sugar is osmotically active, but starch is not). An increase in sugar content would cause water to be drawn into the guard cells, increasing their turgor and opening the stoma. Recent theories of stomatal activity suggest that such alterations in the concentration of osmotically active solutes are probably supplemented by some energy-requiring mechanism.

Conditions of high humidity or dryness may be simulated by mounting small pieces of leaf epidermis in tap or distilled water and in 5% salt (NaCl) solution. Examine guard cells and stomata under high power under these two conditions, and record your observations in the form of sketches. Explain the results.

How do these conditions differ from those to which the plant would normally be exposed?

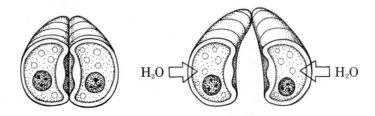

Fig. 4-4. Guard cells. Left: A closed stoma; right: an open stoma.

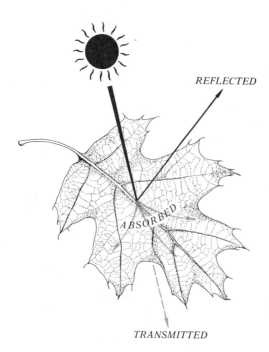

Fig. 4-5. Light striking an object, such as a leaf, may be reflected, absorbed, or transmitted.

PLANT PIGMENTS

When substance absorbs visible light of certain wavelengths (colors) while reflecting and/or transmitting light of other wavelengths (Figure 4-5), we see the substance as colored, the actual color depending upon the light reaching our eyes. Such substances are called **pigments**. The absorption of light energy by a molecule leads momentarily to the formation of an "excited" or high-energy form of the molecule, but in most cases this energy is dissipated without doing any useful work. Plants, however, "trap" the energy absorbed by the chlorophylls and couple this trapping process to the synthesis of carbohydrates. Other plant pigments, such as the carotenoids (e.g., [orange] carotenes and [yellow] xanthophylls), absorb wavelengths of light different from those absorbed by the chlorophylls; this energy is then transmitted to the chlorophylls.

Extraction of pigments with organic solvents

The chlorophylls and carotenoids have been extracted from leaves by your instructor, using hot acetone, an organic solvent in which these pigments are soluble. The extract was then filtered to remove cellular debris.

1. Obtain about 1 cc of the acetone extract in a test tube (about 1-cm level), and add to it about twice its volume of petroleum ether and a few drops of water. Observe that the ether and acetone are mutually insoluble (immiscible) and that the water falls through the upper ether phase and dissolves in the lower acetone phase. Cork the tube, or cover it with a plastic film (such as Parafilm). Keeping your index finger over the top of the tube, invert it several times. This allows droplets of one phase to "fall through" the other so that there is a large surface area of contact between the immiscible solvents. The pigments, or in general any dissolved compounds, wil be "partitioned" (separated) between the phases according to their relative solubility in the two phases. In your experiment, all of the pigments will dis-

solve in the petroleum-ether layer, since the presence of water in the acetone phase makes the pigments practically insoluble in this phase. What does this tell you about the nature of the pigment molecules? Although the carotenoids are also extracted into the ether phase, the ether will appear green because the darker color and greater quantity of the chlorophylls will mask the carotenoids. You will separate these two classes of pigments in the next operation.

2. Decant the petroleum-ether layer into a second test tube, and add 1 cc (20 drops) of 30% potassium hydroxide (KOH) in methyl alcohol. Chlorophyll is soluble in the methanolic KOH, whereas the carotenoids are not. Note the color of each layer after this separation has been accomplished. Record your observations as a sketch of the test tube, showing the different colored layers.

Separation of pigments by thin-layer chromatography

Having just separated the chlorophylls and carotenoids by a simple extraction procedure, you will now separate these pigments by thin-layer chromatography (TLC), a technique more widely used than solvent extractions for separating mixtures of compounds. It is especially useful when the compounds to be separated are very closely related chemically and, therefore, difficult to separate, or when only very small amounts of sample are available. By the methanolic KOH extraction, you were only able to separate the pigments of the leaf into two classes; but by thin-layer chromatography, you will be able to distinguish various pigments within each class. In thin-layer chromatographic separation, the compounds to be separated are spotted along a line near one end of a plastic slide coated with an inert material such as silica or cellulose. After the spot is allowed to dry, the same end of the slide is placed in a closed chamber containing the chosen solvent—petroleum ether:acetone (9:1), in this case.

Chromatography takes advantage of subtle differences in the partitioning or relative solubility of various substances between a stationary phase and a mobile phase. The greater the solubility of a given pigment in the mobile organic phase relative to its solubility in the stationary phase (silica or cellulose), the greater its tendency to remain in the organic phase and the farther its ascent as the solvent rises by capillary action. Subtle differences between the partitioning behavior of various pigments—differences that are often not great enough to permit separation of the pigments by a simple extraction—are greatly accentuated by the relative movement of the two phases. Thus, the different pigments in the original mixture will travel at different rates, and the original spot will separate into a number of different bands. Each band will move a certain fraction of the total distance traversed by the solvent front. This fraction is a characteristic of the particular pigment making up the band and will always be the same whenever this pigment is chromatographed in the same solvent system. If a different mobile phase or a different type of stationary phase were to be used, the distance traveled by each compound would differ from that in the original system. By judicious choice of mobile solvent and supportive medium, it is possible to separate the components of almost any mixture of compounds.

To separate substances other than pigments, the chromatogram must be "developed" in a reagent that differentially colors (stains) the separated substances, or it must be observed under ultraviolet light, where many compounds may be seen either as dark absorption bands or as bright fluorescent bands. One of the most important uses of chromatography is to identify unknown substances. This is done by comparing the mobility of the unknown with reference standards in a number of different solvent systems. If the unknown migrates at the same rate as a given standard in all the solvent systems, then the unknown and the standard are probably identical compounds. You can now appreciate why chromatography has proved itself an exceptionally useful technique for separating and identifying compounds present in very small quantities.

1. At the supply desk, you will find a container of a concentrated extract of leaf pigments in petroleum ether, prepared by your instructor. Obtain a small amount of this extract in a small beaker or flask, and take it to your desk. Before applying the pigment, make a mark on the coated side of the slide about 1.5 cm from the bottom, as shown in Figure 4-6A. With a capillary pipette, apply a spot of pigment to the center of the slide at the level of the mark you just made. Apply the pigments again to the same spot. Repeat this operation as often as necessary (at least five times) to produce a dark green spot, allowing the pigment to dry between applications.

2. In a dry chromatography chamber (Coplin jar) put the solvent (petroleum ether:acetone) to a level of 1 cm.

3. Using forceps to handle the slide, place the slide into the chamber, pigment spot down (Figure 4-6B). Place the lid on the chamber, and wait twenty minutes or more for the pigments to separate. How many pigments were there in the extract? On the basis of color, try to identify them. Remove the chromatogram, allow it to dry, and place it in your laboratory guide. Each student should make a chromatogram for his/her own notebook.

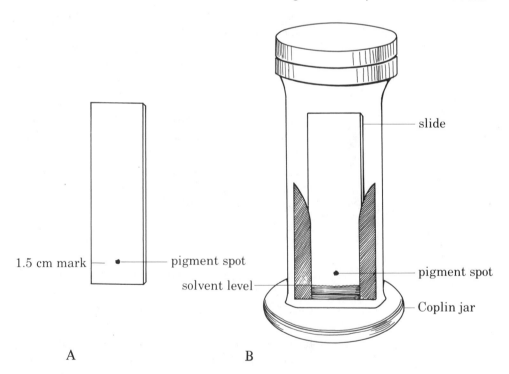

1.5 cm mark — pigment spot

solvent level

slide

pigment spot

Coplin jar

A   B

Fig. 4-6. Thin-layer chromatography. A: Application of pigment to silica-coated slide; B: chromatography chamber with chromatogram in it.

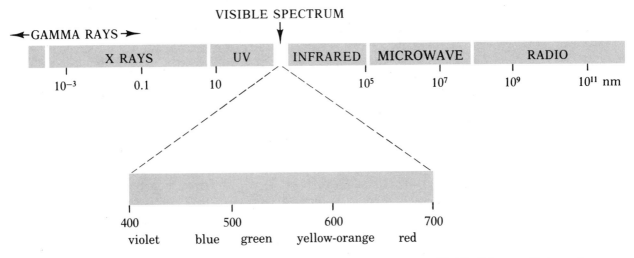

Fig. 4-7. Portion of the electromagnetic spectrum. Visible light constitutes only a very small portion of the total spectrum. Within the visible spectrum, light of different wavelengths stimulates different color sensations in us. Not only vision and photosynthesis, but also other radiation-dependent biological processes, rely on this same small portion of the electromagnetic spectrum (sometimes extended very slightly into the ultraviolet or infrared). Light of wavelengths shorter than about 300 nm (nanometers) is absorbed by the atmosphere, while wavelengths longer than 800 nm have too little energy to drive biological reactions.

Absorption of light by chlorophyll

Visible light, which constitutes a small part of the spectrum of radiant energy (Figure 4-7), can be separated into a variety of colors if passed through a prism. These colors correspond to the different wavelengths of light present in the visible range of the spectrum and are experienced as the colors violet, blue, green, yellow, orange, and red, with the wavelength increasing from violet to red. As mentioned before, a pigment is colored because of its ability to absorb certain wavelengths while reflecting others. In this section, you will study the absorption of light by plant pigments.

A spectroscope (an aimed prism) and a dissecting lamp have been set up at each table in such a way that you may analyze the spectrum of rays transmitted by substances. Between the spectroscope and the lamp, place the following, one at a time:

1. Pieces of colored glass (green, red, blue, yellow, etc.).
2. A container of acetone extract of chlorophyll.
3. A container of acetone.
4. A leaf.

In each instance, note the effect of the spectrum, and record your observations.

Which colors are *absorbed* by chlorophyll? Which are *transmitted*? What wavelengths of light correspond to the colors absorbed by chlorophyll? What wavelengths are known to be maximally effective in photosynthesis? What wavelengths of light are minimally effective? Why was colored glass used? Why was a container of pure acetone used? How does light absorption by the chlorophyll solution differ from that of the leaf? Why should chlorophyll have a different absorption spectrum when in solution? (*Hint:* When chlorophyll is excited in solution, there is no coupling mechanism for using this energy to drive other processes, as there is in the leaf. How do you think the absorbed energy is dissipated in solution? Careful close-range examination of the chlorophyll solution with a light beam shining on it at an angle will help you to find a partial answer to this question. What color do you see?)

MEASUREMENT OF PHOTOSYNTHESIS

While one thinks of a plant as a photosynthetic device, it is important to remember that it respires continuously, as any organism must, in order to obtain energy for its activities. A part of the oxygen released during photosynthesis is required for normal respiration, and when the plant is not carrying out photosynthesis, it must obtain respiratory oxygen from the environment. The consumption of oxygen in the dark may be used as a measure of respiration. In the light, with both photosynthesis and respiration proceeding simultaneously, the oxygen evolved is the difference between these opposing reactions. Under optimal conditions, the rate of oxygen production by photosynthesis is ten to twenty times greater than the rate of oxygen consumption by respiration. Hence, under such conditions, the oxygen *measured* during photosynthesis represents 90% to 95% of the total oxygen produced.

Effect of light intensity on oxygen production

1. Fill a test tube to within 5 mm from the top with 2.5% sodium-bicarbonate solution. Sodium bicarbonate serves as a carbon-dioxide source, according to the following reaction:

$$NaHCO_3 \xmid{H_2O} NaOH + CO_2$$

Place a healthy sprig of *Elodea,* about 10 cm to 15 cm in length, into the tube with the cut end up. Before completely immersing the *Elodea,* cut 3 mm to 4 mm off the cut end of the stem.

2. Fill a battery jar or other flat bottle with cold water, and place it between the test tube and a 150-watt lamp to act as a heat filter. This will prevent large temperature fluctuations in the test tube.

3. Place the tube 25 cm from the light source, and allow the tube to sit for five minutes to equilibrate to room temperature.

4. Count the number of oxygen bubbles evolved over a two-minute period. Record your data in Table 1. Repeat the procedure again, and determine the average number of oxygen bubbles evolved per minute.

5. Move the test tube 50 cm from the light source. Allow the tube to stand for five minutes, then count the number of bubbles evolved. Repeat the procedure at 75 cm and 100 cm.

Plot the data on Graph A. Be prepared to interpret and explain your data. Does the oxygen evolved represent the *total* oxygen produced in photosynthesis? What factors influenced the rate of photosynthesis in your experiment? Design experiments to demonstrate how the rate of photosynthesis is influenced by each of the following: light quality, temperature, and concentration of carbon dioxide. If there is time, your instructor may be able to provide materials so that you can carry out some of the experiments you designed above.

DISTANCE FROM LIGHT (cm)	NUMBER O$_2$ BUBBLES EVOLVED/2 MIN. READING 1	READING 2	AVERAGE NO. BUBBLES/MIN.

TABLE 1.   NUMBER OF OXYGEN BUBBLES EVOLVED PER MINUTE.

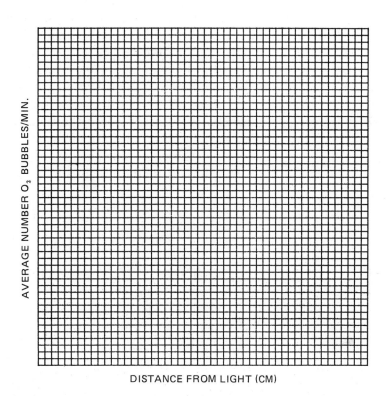

GRAPH A.   O$_2$ PRODUCTION IN PHOTOSYNTHESIS.

Light-dependent Reaction   The chemical reactions that occur during photosynthesis are divided into two groups: light-dependent reactions and dark reactions. During the light-dependent reactions, energy from the sun (in the form of protons) raises the energy level of an electron in chlorophyll, which in turn passes through an electron transport system in the thylakoid membrane. Energy liberated in the electron transport chain is used to pump H$^+$ ions across the thylakoid membrane; NADPH is produced from NADP, and O$_2$ is liberated as part of the same series of reactions in the electron transport chain. The energy of the H$^+$ gradient across the thylakoid membrane is then used to produce ATP. The dark reactions use the ATP and NADPH to synthesize carbohydrates.

In this part of the experiment, the necessity of light and chlorophyll for the production of NADPH will be investigated. If, in the presence of chlorophyll and light, electrons can be removed from water and then transferred to electron-acceptor molecules, it is possible to test whether this reaction has occurred by using a special dye. When this dye is added to the test tube, the dye is blue; however, when the dye is reduced, it becomes colorless, indicating that NADPH has been produced.

Demonstration of the light-dependent reaction

Obtain five test tubes from the supply desk. Cover the outside of one of the test tubes with aluminum foil, one with a single layer of green cellophane, and one with a single layer of red cellophane. The remaining two test tubes should not be covered.

A chloroplast suspension from spinach leaves has been prepared by the instructor. Under the compound microscope, examine a drop of the chloroplast solution on a slide to be certain that you have intact chloroplasts.

To each of the test tubes you prepared, add 2 ml of the chloroplast suspension. Add 4 drops of the dye 2,6-dichlorophenolindophenol (an electron acceptor) to all tubes except one of the uncovered tubes, and mix. The tube without the dye will serve as the control for the original color of the chloroplast suspension. Place the tubes in a rack 50 cm from a 100-watt light. Examine the tubes after ten minutes in the light. Record the color change in each tube, and explain the results.

Which wavelengths of light are most effective?

QUESTIONS FOR FURTHER THOUGHT

1. What physical characteristics make the leaf an effective photosynthetic organ?

2. What special adaptations does the leaf have for gas exchange?

3. Suggest a hypothesis to account for the fact that leaves of different plants have different shapes.

4. What is the role of carotenoids in photosynthesis?

5. What colors of the visible spectrum are absorbed by the leaf?

6. Explain how the level of carbon dioxide can exert control on the light-dependent reaction.

HETEROTROPHIC NUTRITION

Biological Science, 4th ed.
Chap. 10, pp. 238–271
Elements of Biological Science, 3rd ed.
Chap. 9, pp. 151–170

Heterotrophic organisms, you may recall, are incapable of synthesizing their own organic compounds from simple inorganic nutrients and must, therefore, rely largely upon prefabricated, high-energy organic molecules. Since the majority of such molecules are too large and insoluble to pass through semipermeable membranes, they must be broken down into their smaller, absorbable subunits.

We know that **digestion** is the breakdown of nutrients within our own digestive tracts; but we must also remember that the term describes *any* breakdown of a complex compound, such as protein, fat, starch, or compound sugar, into its constituent building blocks (amino acids, fatty acids, glycerol, and simple sugars), and that such processes occur regularly *within* most cells of both heterotrophs and autotrophs in the course of normal cellular metabolism. However, before heterotrophs are able to absorb organic nutrients into the surrounding cells, they must first ingest their food into a specialized digestive structure where enzymatic digestion takes place.

In the laboratory today, you will examine the digestive system of a heterotrophic animal and observe some of the digestive adaptations for the processing of organic nutrients prior to absorption.

INTRODUCTION TO DISSECTION

Inasmuch as this is your first dissection exercise, we will begin by learning how to use the dissecting tools and by reviewing some procedures that will be followed in this and other exercises.

Tools and General Procedures   Examine the contents of your dissecting kit. A **scalpel,** or a single-edge razor blade, is used to make initial cuts through epidermal layers, and on rare occasions, when a broad cutting edge is required, it may be used to incise or excise some organ. However, the scalpel or razor blade should *never* be used in routine dissection; if you have formed the habit of using it frequently, you should break that habit at once. **Scissors** are the primary cutting instruments, but they may also be used for spreading tissues by inserting the closed points and then opening them to push the tissues to either side. The **forceps** is used as a probe, as a manipulator, and as a means of lifting structures to be cut so that underlying areas are not damaged. The **probe** is used for lifting, pushing aside, feeling (where a structure may not be clearly visible), and in tracing the course of ducts. It is also effective in tearing loose bits of connective tissue. The **teasing needle** is useful for making careful punctures or on the few occasions when a sharp-pointed instrument is necessary. Be sure to keep your instruments clean and dry, and they will remain serviceable for a long time.

In carrying out your dissections, you should carefully follow the directions in this guide and those instructions given by your instructor. Dissection is to be undertaken with great care and is a procedure to help you locate, expose, and learn the anatomy of an animal. It is *not* a random cutting-up of a carcass. Always remember that you should do as little cutting as possible because the more you cut, the more you will alter the structural relationships among the parts. It is your objective always to learn the *original* organizational patterns, not some new arrangements created as artifacts of your dissection. You will have to cut out connective tissues in order to see the relationships between organs, but you should always know *what* you are cutting and *why* you are cutting. If you keep this in mind, you should have no difficulty avoiding damage to your specimen and preparing a good dissection.

You are encouraged to compare notes with others as you work, but the actual dissection on the specimen should be your own. Since you will need to have the specimen in adequate condition for future study, work with care. When you have finished today, place your animal back in its plastic bag, tightly close the bag, tie the bag, and attach to the bag a tag on which your name and laboratory section appear. Follow your instructor's directions for cleaning up and disposing of waste tissue.

Anatomical Terms   In order to utilize fully the directions given in this and succeeding laboratories, you will need to learn a few adjectives used to denote orientation and direction in anatomical work. Be sure you understand the following, and refer back to this list whenever necessary:

right and left	— the *rat's* right and left
anterior	— toward the head
posterior	— toward the tail
caudal	— toward the tail
dorsal	— toward the back; up
ventral	— toward the belly; down
lateral	— toward the side (right or left)
medial	— toward the middle or center
proximal	— near a specified point of reference
distal	— away from a specified point of reference
pectoral	— referring to the shoulder region
pelvic	— referring to the hip region

VERTEBRATE DIGESTION

Of the several classes of vertebrates, the one in which we are most naturally interested is that to which we ourselves belong: Mammalia, or mammals. The features that distinguish mammals from other vertebrates are (1) the presence of hair sometime during the life cycle and (2) the existence of mammary glands in the female. The mammal that you are about to study is the rat (Figure 5-1). Using Appendix 2, classify the rat before you proceed.

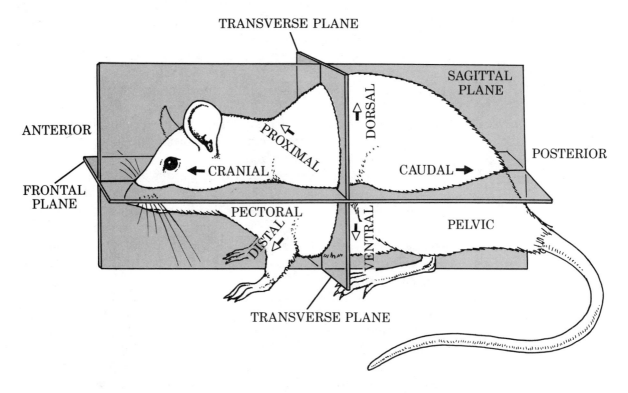

Fig. 5-1. Lateral view of the white rat, illustrating the planes and directions used in anatomical terminology.

Since this is the first time you have examined the rat, we will not only be concerned with the digestive tract, but will also look briefly at some external characteristics of the animal. Likewise, as the dissection proceeds, your attention will sometimes be called to internal structures not directly related to the alimentary canal, but more easily examined now than at a later stage in your work.

Examining External Features of the Rat

Body regions   The body of a mammal is arbitrarily divided into a series of readily identifiable regions: head and neck, trunk, and tail. The trunk consists of an anterior thorax (protected by the ribs) and a posterior abdomen. Note that the rat has hair covering all body regions except the tail. The hair functions mainly as a temperature regulator. A layer of air gets trapped between the skin and the hair, and serves as insulation. This layer of insulation is regulated by raising or lowering the hair. Unlike humans, rats have no sweat glands on their body, only on the pads of their feet. When the animal is overheated, water is evaporated from its mouth by panting. What body coverings do other classes of vertebrates such as fish, amphibians, reptiles, and birds possess?

The head   Locate the **eyes, nostrils, mouth,** and external ears or **pinnae** (singular, "pinna"). The eyes are surrounded by an upper and lower eyelid plus a reduced eyelid, the nictitating membrane. The pinnae direct sound waves to the eardrum. The rat, like most terrestrial mammals, possesses an acute sense of vision, smell, and hearing.

The long whiskers around the snout are called **vibrissae**. These are special sensory hairs that are extremely sensitive and of great importance to a nocturnal animal. Where are additional sensory hairs located?

The trunk   In the anterior portion of the trunk, feel the sternum (breastbone) and the ribs, which form a thoracic cage for the heart and lungs. The abdomen contains most of the digestive tract.

Examine the four legs. Feel their various joints and parts, referring to Figure 5-2 when necessary, and determine which parts of the forelimb correspond to your upper arm, elbow, lower arm, wrist, and hand. Do the same for the hind limb, comparing it to your own leg. How many digits are there on each foot? Note the pads and claws on the forefeet and hindfeet. The rat walks on the digits with the rest of the foot elevated, while humans walk on the entire sole of the foot.

You will now determine the sex of the animal you are examining. The female has three exterior openings: (1) the anus, located just under the tail, is the posterior opening of the digestive tract; (2) the vagina, located just anterior to the anus; and (3) the urethral opening, anterior to the vagina. Males possess a scrotum, containing the testes, anterior to the anus, and, in front of the scrotum, a single urogenital opening through which the penis is everted during copulation.

If your specimen is a female, determine the location of the nipples and the number of pairs. Nipples are the external openings from the mammary glands, a distinctive feature of mammals. Compare the number of nipples found on your animal with that of your classmates' animals. Does the number vary?

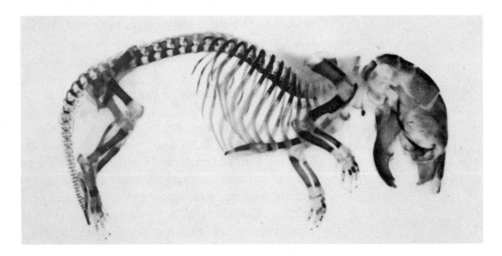

Fig. 5-2. Rat cleared to show developing bones. *(Courtesy James Hanken, University of Colorado, Boulder.)*

The tail   Examine the tail carefully. Note that the tail is covered by scales with small clusters of hairs between the scales. From which group of vertebrates do you think the mammals evolved?

The Digestive Tract   Animal digestive tracts vary widely, and the differences are often closely correlated with the type of food consumed. You will examine the rat's digestive tract in detail, but you should note other animals on demonstration.

The oral cavity   With heavy scissors, cut posteriorly from one corner of the mouth, through the tissue and bone, to the base of the tongue. Using Figure 5-3 as a guide, follow the dotted line from the corner of the mouth to the bottom of the ear. (*Note:* For this and all subsequent cuts, follow the dotted lines in Figure 5-3. Do not be afraid to cut through the rat's jawbone—the cut is necessary to expose deeper parts of the oral cavity.) Repeat this operation on the other side, and spread the jaws so that the oral cavity is exposed to view. Open the **jaws** as widely as you can without cutting. Identify the **oral cavity**, with the **tongue** on its ventral surface and the **hard palate** forming its dorsal surface. The hard palate separates the oral cavity from the **nasal cavities**.

Examine the tongue carefully, and look at the bones on a skeleton in the region where the tongue would be attached. How is it attached? What is the role of the tongue in feeding? Can you think of any animals in which the tongue is adapted for other functions?

Locate the papillae scattered over the surface of the tongue. These contain taste buds.

Examine the hard palate in more detail, and locate its posterior border. Posterior to the hard palate is the **soft palate**, which contains no bone. In humans, a small posterior extension of the soft palate, the **uvula**, hangs down into the throat.

At the anterior end of the oral cavity, locate the four **incisor teeth**, which are specialized for gnawing—a characteristic activity of rodents. The incisors grow continually during the animal's lifetime and are kept sharp by the grinding of the upper teeth against the lower teeth. The three pairs of **molar teeth**, which grind up the larger pieces of food, are located toward the back of the oral cavity.

The most posterior part of the oral cavity is known as the **pharynx**. This cavity, a common passageway for the food and respiratory canals, is delimited ventrally by the caudal base of the tongue and dorsally by the caudal border of the soft palate. (Be sure that you have cut far enough into the jaw, or you will not be able to spread the jaws properly to see the pharynx.) Note the opening into the pharynx of the **nasopharynx**; food and respiratory canals cross in the pharynx (Figure 5-4).

Identify the **epiglottis**, a median flap at the base of the tongue. The epiglottis partially covers an opening, the **glottis**, which is the entrance into the larynx and trachea. When swallowing food, the larynx is raised forward slightly and the epiglottis moves back over the glottis, preventing food from entering the larynx. Posterior and dorsal to the glottis, find the opening into the **espohagus**, the tube of the digestive tract that leads through the neck and

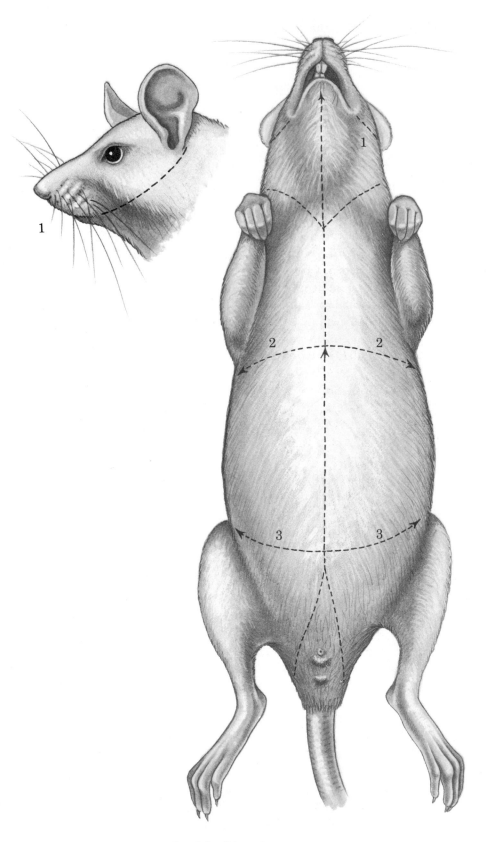

Fig. 5-3. Dissection cuts.

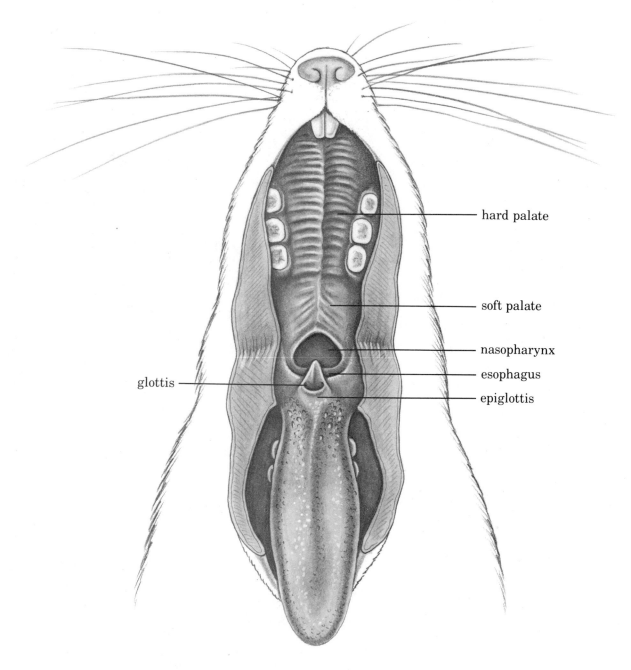

Fig. 5-4. Rat—oral view.

thorax to join the stomach (Figure 5-5). Pass the end of your probe into the glottis and then into the esophagus; be certain that you can distinguish between these two passageways.

The neck region   Place the rat ventral side up. Make a small incision through the skin of the abdomen along the mid-ventral line. Holding the skin with a forceps, continue cutting upward toward the lower jaw. Turn the animal around, and

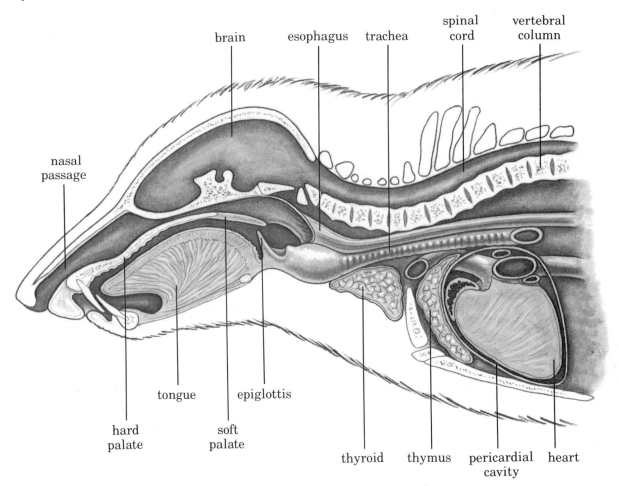

Fig. 5-5. Rat head, neck, and chest—sagittal section.

continue cutting along the mid-ventral line around both sides of the genital openings to the anus. Make cuts laterally through the skin, as shown in Figure 5-3, being careful not to cut the organs underneath.

Take two pieces of soft cord, and tie them to the ankles of each leg in such a way that the animal lies in the dissecting pan ventral side up, with the strings under the pan (Figure 5-3); this will hold the specimen in position. At the end of the period, slip the strings off the ends of the pan, leaving them attached to the animal so that you can position it quickly at the beginning of your next dissection.

Using a blunt probe, separate the skin in the neck region from muscle and other tissues. Note the large paired **submaxillary salivary glands** (Figure 5-6). What is the function of these glands? Examine the area carefully for the presence of small pink glandular structures, which are the lymph nodes. Bend back the submaxillary glands. Probe down into the deeper layers of the neck. Medially, beneath several strips of muscle, will be found the hard-walled **larynx** and **trachea**, parts of the respiratory passage that will be examined later in more detail. In order to see the trachea and larynx clearly, it may be necessary to cut the longitudinal muscles where they insert on the pectoral girdle. On the ventral surface of the trachea, find the **thyroid**

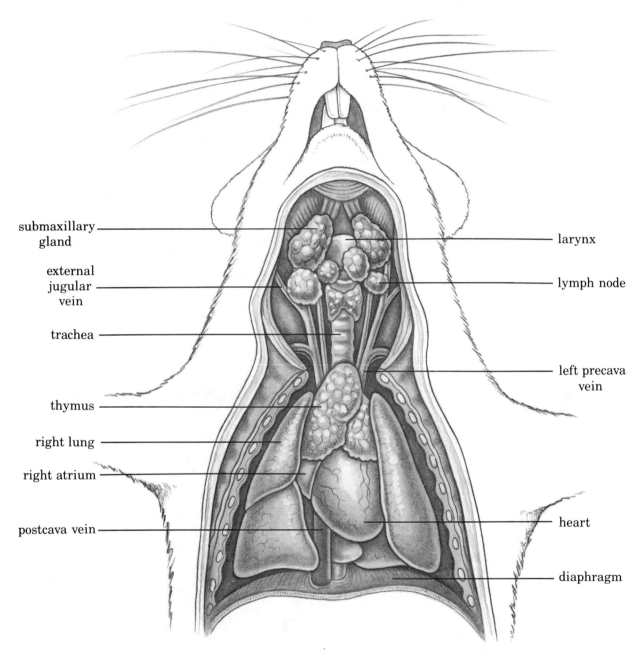

Fig. 5-6. Rat neck and chest—ventral dissection.

gland. This is a small, brown, H-shaped endocrine gland that secretes thyroxin, a hormone that regulates metabolism.

Dorsal to the trachea, probe for the **esophagus,** a thin-walled, narrow tube. One easy way to be certain that you have located the esophagus is to open the mouth, probe well down into the esophagus, and then feel in the neck for the tube containing the hard probe. Be certain that you understand the relationships of the food and air passages in the head and neck.

The thoracic region   The esophagus continues posteriorly through the cavity of the thorax without changing significantly. We shall, therefore, not open the thorax at this time, but shall examine its contents later in connection with the circulatory and respiratory systems.

Examining the abdominal region

The abdominal region is that area, posterior to the ribs, in which the ventral body wall has no bony support. The body wall in this region encloses a large **peritoneal cavity**, lined by epithelial tissue called the peritoneum. Most of the digestive, excretory, and reproductive organs are found within the peritoneal cavity, together with some organs assigned to other systems. Each of the three systems mentioned above has one or more ducts that pass to the exterior through a ring of bones called the **pelvic girdle**. Examine this region on a skeleton (Figure 5-7).

Pin back the lateral flaps at the side of the abdomen.

The **liver**, which is the largest organ in the abdomen, is made up of four lobes. Note that the anterior surface of the liver is smoothly convex and fits snugly into the concavity of the **diaphragm**, a muscular partition that separates thoracic and abdominal cavities. Push the liver aside with your fingers, and examine the diaphragm. (*Note:* If the peritoneal cavity of your specimen is partially filled with dark, brownish material, take your animal to the sink

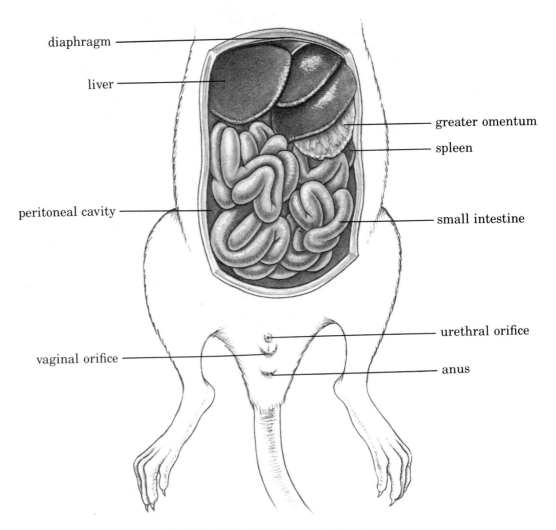

Fig. 5-7. Abdominal viscera of a female rat.

and rinse it thoroughly before you continue to dissect. This material is clotted blood which filled the peritoneal cavity when one or more of the blood vessels associated with abdominal organs burst during injection of the circulatory system. It does not harm the specimen or make it less useful for dissection, but it should be thoroughly cleaned out.)

Push the liver aside, and identify the **stomach**, a large sac dorsal to the liver on the left side. Locate the point at the anterior end of the stomach (near the mid-line of the body) where the esophagus penetrates the diaphragm and then almost immediately joins the stomach. At its posterior end, the stomach makes a curve to the right and narrows to join the anterior end of the **small intestine**. The constriction at the junction between the stomach and the small intestine is called the **pylorus**. A long, flat, reddish organ, the **spleen**, which is not part of the digestive system, is attached to the stomach by a mesentery called the **greater omentum**. The greater omentum also functions in the storing of fat.

What is the function of the spleen? What digestive functions take place in the stomach? Does any absorption take place in the stomach? How is the stomach related to the phenomenon of discontinuous feeding? How would your own eating habits have to be modified if your stomach were removed?

Examine more closely the anterior end of the small intestine, known as the **duodenum**. The liver and the duodenum are connected in the pyloric region by the **bile duct**, a long duct that extends through the **lesser omentum** (mesentery) stretching between the two organs. Under a dissecting microscope, the bile duct may be traced into the liver. In other vertebrates, the bile duct would also branch to the **gall bladder**, found on the underside of the liver. However, the rat does not have a gall bladder; therefore, bile is not stored in this organ.

How does the liver aid in digestion? What are some of the liver's non-digestive functions?

Lift the stomach, and locate the **pancreas**, a light-colored, diffuse gland lying in the mesentery between the stomach and the small intestine. The pancreas has a duct that empties into the small intestine near the pylorus.

What enzymes are secreted by the pancreas, and what is their action? How else does the pancreas function?

With a scalpel or razor blade, make an incision into the posterior end of the stomach, and carry the cut through the pylorus a short distance into the duodenum. Find the **pyloric sphincter** muscle.

What is the function of this muscle?

From the pyloric region, the small intestine runs posteriorly for a short distance and is then thrown into an irregular mass of bends and coils held together by a common mesentery containing numerous blood vessels.

An exceedingly common method of increasing the effective digestive and absorptive area in the intestine is the presence of fingerlike extensions of the inner layer of the small intestine. These projections are called **villi** (singular, "villus"). Open the small intestine of the rat, and examine the villi under a dissecting microscope.

What enzymes are secreted by the small intestine, and what is their action? What is the chief function of the large intestine? In what animals might this function be relatively unimportant?

The distal end of the small intestine joins the **large intestine** posteriorly on the right side of the abdominal cavity. A blind sac, the **caecum**, will be found projecting from the large intestine near the point of juncture. In human beings, there is a smaller end to the caecum. What is it called?

Follow the main portion of the large intestine, known also as the **colon,** as it runs from the point of juncture with the small intestine anteriorly, then turns and runs posteriorly along the mid-line of the dorsal wall of the abdominal cavity. In the pelvic region, the digestive canal is called the **rectum**; it will be seen more clearly later. The bulblike objects within the rectum are fecal pellets containing undigested material. Posteriorly the digestive tract opens to the outside through the **anus.**

Digestive-system modifications

One major function of teeth is mechanical breakup of large food masses, which provides a greater surface area upon which the enzymes involved in chemical digestive processes can act. Animals have developed many other means of mechanical breakup. Examine the demonstration specimens of bird, cockroach, earthworm, and crayfish, showing the organs these animals have developed for physical breakup of food.

Also observe the variations in teeth in these specimens; for each, try to correlate the differences with the mode of feeding and the type of food utilized. Mechanical breakup is only one of the many functions of teeth. What are some of the other functions?

Examine the jaws on a demonstration skeleton of a mammal, and note how they fit one another. Are both jaws movable?

We know from study of fossils that the first vertebrates lacked jaws. These early vertebrates were sluggish, filter-feeding organisms that lived along the bottom of ponds and streams. Early descendants of these jawless vertebrates developed a predatory mode of life and evolved jaws for seizing and holding prey. From this stock eventually evolved fish, amphibians, reptiles, birds, and mammals, all of which have retained the jaws in a more or less modified form. On demonstration is a **lamprey,** one of the few extant members of the ancient class of jawless vertebrates. Though jawless, the lamprey itself is modified from the early vertebrates in that it abandons the mud-grubber mode of existence as an adult, but remains buried in the mud throughout its larval stages.

QUESTIONS FOR FURTHER THOUGHT

1. What are the digestive functions of the liver?

2. The liver has many metabolic functions in addition to those concerned directly with digestion. Can you suggest adaptive advantages to having so many important functions localized in a single organ? Can you think of any disadvantages in such an arrangement?

3. Discuss and compare digestive adaptations in herbivores and carnivores.

4. Many animals have gizzards. What do the diets of these animals have in common?

5. Trace a molecule of an absorbable nutrient and a molecule of a non-absorbable nutrient through the digestive system, naming all structures involved in the passage from ingestion to absorption or defecation.

6. Explain the function of the gall bladder in mammals.

7. How is an animal's size related to its metabolic rate?

TOPIC **6**   GAS EXCHANGE
IN ANIMALS

Biological Science, 4th ed.
Chap. 11, pp. 280–294
Elements of Biological Science, 3rd ed.
Chap. 10, pp. 176–184

We have seen how high-energy organic compounds are made by autotrophic organisms and have observed some of the adaptations made for processing organic molecules by heterotrophic organisms. The largest amount of energy (in the form of ATP) can be obtained from these high-energy organic compounds if they undergo complete cellular respiration; for this to occur, oxygen is usually necessary. Cells must, therefore, have a continuous supply of oxygen and a means of releasing carbon dioxide. Since exchange of gaseous raw materials and waste products of metabolism is a basic problem facing all organisms, it is not surprising to find that special organs have evolved, adapted especially for carrying on the exchange of these materials with the environment.

We must remember that, ultimately, the critical process in gas exchange is the diffusion of gases into and out of individual cells. Such diffusion is possible only across a moist membrane. Therefore, in order to have sufficient gas exchange with its environment, an organism must have a surface of moist cell membranes exposed to the environment, and exposed in such a way that the chances of drying out are minimized. Another requirement is a respiratory surface large enough, relative to the size and activity of the organism, to allow for an exchange of sufficient magnitude to meet the organism's needs. Since such a large, thin, moist surface is often very fragile and easily suffers mechanical damage, there has generally been selective pressure for the evolution of protective devices, especially when the respiratory surface is an evaginated one.

Respiratory surfaces (excluding very small organisms in which the whole body surface is a respiratory surface) may be grouped into two general categories: **invaginations** (inpocketings) of the body surface, and **evaginations** (outpocketings). Which of the two would you expect to find more commonly among terrestrial organisms? Why?

In this exercise, the respiratory system of a mammal will be examined in detail. While studying this system, keep in mind that this is one way of solving the following problems that all organisms must meet:

1. The need for a respiratory surface of adequate dimensions.
2. The need for keeping the surface moist.
3. The need for protecting the surface from mechanical injury.
4. In the case of many multicellular organisms, the need for a method of transporting gases between the sites of exchange with the environment and the more internal body cells.

51

RESPIRATORY SYSTEM OF A MAMMAL

With the exception of certain forms that have secondarily returned to the water, mammals have become entirely emancipated from the aquatic environment, and they all carry out gas exchange by means of lungs. All mammals depend upon a blood circulatory system to transport gases between the moist respiratory surface of the lungs and the individual body cells.

Examining the rat's respiratory system

1. On your rat, once again locate the nostrils and the inner end of the nasal passages, just behind the soft palate. To examine the nasal passages more carefully, use a scalpel or heavy scissors to cut across the nose about 1 cm back from the nostrils. Removal of the tip of the nose will expose the nasal passages, separated from each other by a bony and cartilaginous partition called the nasal septum. The floor of the passages is formed by the hard palate anteriorly and the soft palate posteriorly. Along the lateral walls of the passages are large folds, which cause eddies in inspired air and thereby serve to increase the contact of the air with the extensive epithelial surface within the nasal passages. Why is such contact important?

2. Review your understanding of the **pharynx** as a common passageway for air and food. Also review the location of the **glottis, epiglottis, trachea,** and **esophagus.**

3. Probe into the dissection already made in the neck, and carefully clean all muscle tissue from the surface of the hard-walled **larynx.** (The walls are hard because they contain cartilage.) The larynx is the chamber in which the anterior opening, the glottis, was seen on the ventral side of the pharynx. It is the anterior portion of the trachea. Slit open the larynx along the mid-ventral line, and locate the **vocal cords,** a pair of dorsoventral folds of tissue in the lateral walls.

Posteriorly, the larynx connects with the **trachea,** identifiable as a tube whose walls are supported by numerous rings of cartilage. Clean and examine it carefully. Do the cartilage rings extend all the way around the trachea? What is their function? Below the larynx, on the ventral surface of the trachea, locate the **thyroid gland.** It appears much darker in color than other organs in this region. The thyroid is a part of the endocrine system, not the respiratory system, but it is most easily located at this time.

4. Locate the posterior boundary of the thoracic cavity by reaching through the incision with a finger and feeling for the **diaphragm.** Begin your dissection from the previous mid-ventral incision near the diaphragm. Using a pair of scissors, cut through the rib cage and body wall, slightly to the left of the mid-ventral line (Figure 6-1). Raise the body wall as you cut so that you do not damage the heart and lungs. Continue cutting up through the center of the clavicle until you can find the region where the trachea and the esophagus enter the thorax. Fold back the chest wall, exposing the thoracic cavity (Figure 6-2). It will be necessary to break the ribs in order to do this. In folding back the right flap, you will need to tear the thin membranes that divide the thoracic cavity into three compartments: the **left pleural cavity,** containing the left lung; the **right pleural cavity;** and the **pericardial cavity,** containing the heart.

5. Examine the lungs; locate the four lobes of the right lung and the single lobe of the left lung. The fourth lobe of the right lung lies dorsal to the large vein entering the heart, the vena cava (Figure 6-3). Remove a thin

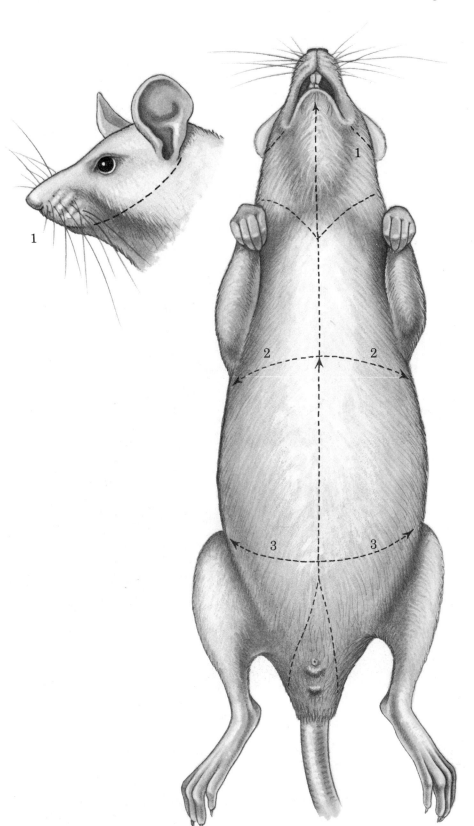

Fig. 6-1. Dissection cuts.

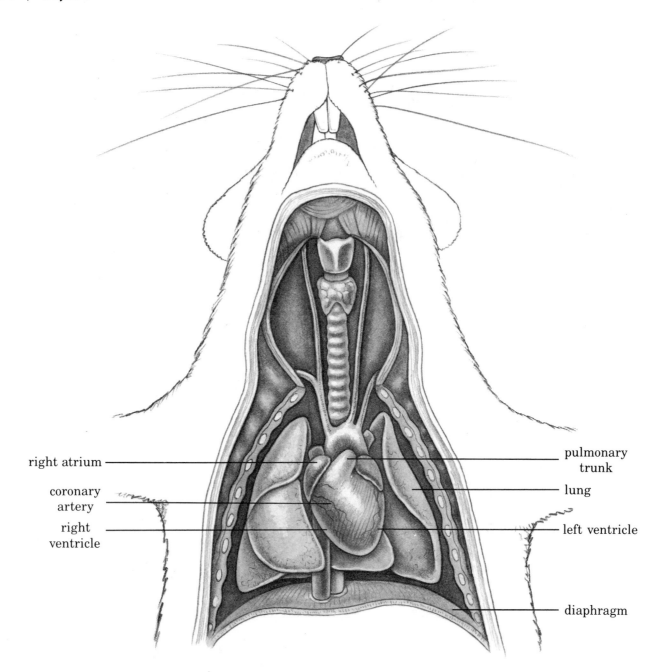

Fig. 6-2. Rat thoracic cavity—ventral dissection.

right atrium

coronary
artery

right
ventricle

pulmonary
trunk

lung

left ventricle

diaphragm

slice of the lung, place it on a microscope slide, and observe its spongy appearance.

6. You will be unable to examine the division of the **trachea** into the two **bronchi** because these tubes are dorsal to the heart and its attached vessels, which we will examine in detail later. At that time, the elements of the circulatory system will be removed, exposing the bronchi to view.

The bronchi subdivide repeatedly, forming **bronchioles**. The ultimate subdivisions terminate in blind sacs called **alveoli** (singular, "alveolus"). The alveoli are the real functional units of the lung; it is across their moist walls

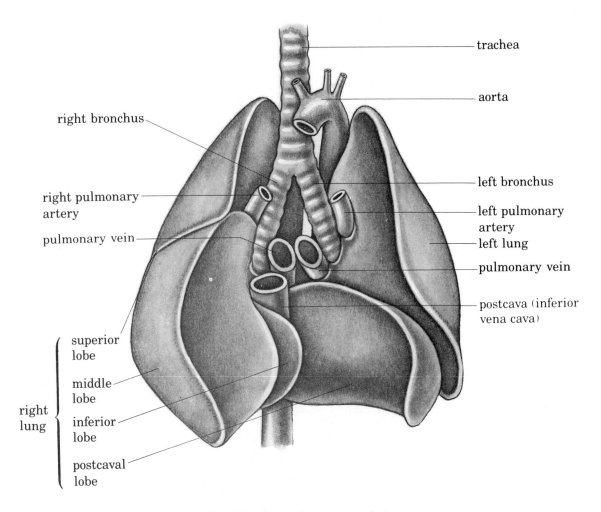

Fig. 6-3. Lungs of a rat—ventral view.

that gas exchange actually takes place. Figure 6-4 shows the relationship in humans between the air-transporting tubes and the alveoli.

7. Examine the diaphragm, and identify the central tendon, which is a membranous area in the diaphragm's center. Muscles, arranged radially around the rim of the diaphragm, attach it to the chest wall. The chest wall itself has an outer layer of skin, a middle layer of ribs and rib muscles, and an inner, thin layer of tissue lining it, the parietal pleura. The surface of the lungs is covered by the visceral pleura. Similar membranes cover the heart.

Be sure you thoroughly understand the roles of the rib cage and the diaphragm in breathing. At the demonstration desk, a model has been set to help you understand the concept of **negative pressure breathing**, as performed by mammals.

DETERMINATION OF LUNG CAPACITIES

Breathing is a cyclic process involving inspiration and expiration. The volume of the air inhaled and exhaled at each breath can be measured with an appara-

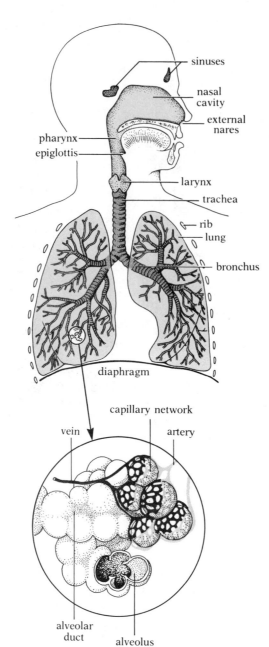

Fig. 6-4. The human respiratory system. The enlargement shows a few sectioned alveoli and other alveoli surrounded by blood vessels. Actually, all alveoli, including those lying along alveolar ducts, are surrounded by networks of capillaries.

tus called a **spirometer** (Figure 6-5). These measurements (Figure 6-6) can tell us whether or not lungs are functioning normally, as influenced by sex, age, and weight. The following respiratory volumes will be determined:

Measuring volumes and capacities

1. **Tidal Volume** (TV). The amount of air inhaled or exhaled during normal quiet respiration in called tidal volume. The average normal tidal volume is 500 ml; however, about 150 ml consist of air in the conducting tubes of the respiratory system and are not available for gas exchange. Determine your tidal volume in the following manner:

    a. Breathe normally for one minute.

    b. Inhale a normal breath through the nose; then exhale normally into the spirometer mouthpiece.

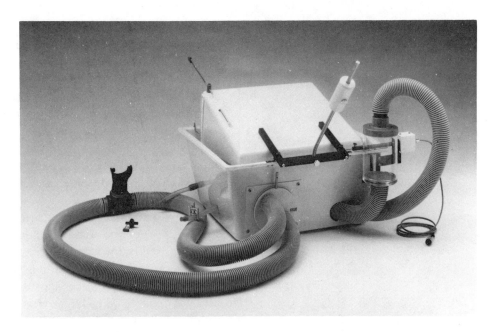

Fig. 6-5. A recording spirometer. *(Photo: Courtesy Harvard/SRI.)*

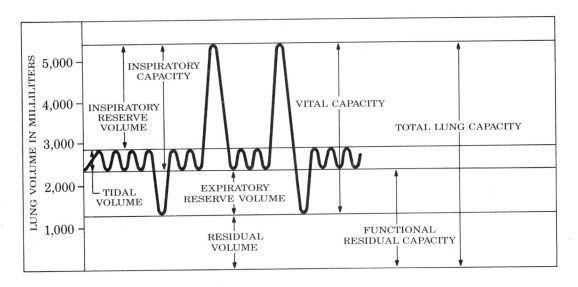

Fig. 6-6. Spirogram of lung capacities.

    c. Repeat this procedure two times.

    d. Record your results.

   2. **Expiratory Reserve Volume** (ERV). The expiratory reserve volume is the amount of air that remains in the lungs after normal exhalation. This volume of air is about 1,000 ml.

    a. Breathe normally for one minute.

    b. Exhale normally (*not* into the spirometer); then exhale all the air possible from your lungs into the spirometer mouthpiece.

    c. Repeat this measurement twice.

3. **Inspiratory Reserve Volume (IRV).** This is the amount of air that can still be inhaled into the lungs after a normal inhalation. This volume varies a great deal among individuals and may range from 1,000 to 1,500 ml.

    a. Breathe normally for one minute.

    b. Inhale as deeply as possible; then exhale normally into the spirometer mouthpiece (do not forcibly exhale). From this value, subtract the TV to obtain the IRV.

    c. Repeat twice.

4. **Vital Capacity (VC).** The vital capacity of the lungs is the measure of the combined tidal, expiratory reserve, and inspiratory reserve volumes of air.

    a. Breathe slowly and deeply two or three times.

    b. Breathe in as deeply as possible; then exhale as much air as possible slowly and forcibly into the spirometer mouthpiece.

    c. Repeat several times. Your readings should be within 100 ml each time. Compare your vital capacity with those of your classmates. Do athletes tend to have larger vital capacities? Your result should approximate the added totals of TV + ERV + IRV.

5. **Inspiratory Capacity (IC).** The maximum volume of air that can be inhaled after normal exhalation is called the inspiratory capacity. This volume is calculated by the formula:

$$IC = VC - ERV$$

6. **Residual Volume (RV).** There always remains a volume of approximately 1,200 ml of air in the lungs that cannot be expired even forcibly, and this is referred to as the residual volume. This volume is determined by complex methods and cannot be measured by the spirometer.

MEASUREMENT OF RESPIRATION

In order to understand the importance of oxygen in cellular respiration, one needs to measure the amount of oxygen consumed by an organism in a given amount of time. To measure respiration, we shall use a respirometer (Figure 6-7).

The oxidation of glucose may be summarized by the equation:

$$C_6H_{12}O_6 + 6O_2 \longrightarrow 6CO_2 + 6H_2O + energy$$

As the equation shows, the quantity of oxygen consumed is equal to the quantity of carbon dioxide released. Since carbon dioxide, like oxygen, is a gaseous component of air, the volume of air in the experimental tube of the respirometer will remain constant unless the carbon dioxide produced by respiring cells is absorbed or in some manner retained in solution. A granular soda lime (solid sodium hydroxide and calcium hydroxide) is, therefore, added to the system to absorb carbon dioxide.

Measuring respiration in insects

The measurements will be made on large insects. We use insects for two reasons: their size and their relatively high oxygen consumption for animals of their weight. Terrestrial arthropods such as insects, centipeds, millipeds, etc., have an invaginated gas-exchange system known as a **tracheal system.** Air enters through **spiracles,** small openings in the body wall, and is carried

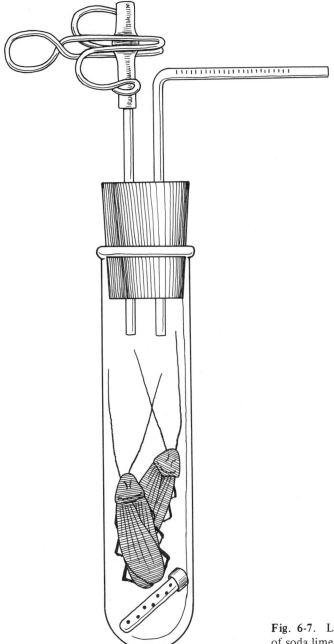

Fig. 6-7. Large insects and container of soda lime in experimental tube.

from the spiracles by a system of branching tubules called **tracheae.** The tracheae usually branch repeatedly until the finest tubules reach individual cells or groups of cells. At the very end of the tracheae, then, are the moist cell membranes across which respiratory gas exchange occurs. In the tracheal system, the enormous surface area required is usually achieved by repeated subdivision of the air tubes, not by convolution of surface and branching of tubes within a restricted structure such as a lung.

1. The respirometer (experimental tube) should be assembled with a perforated container of soda lime at the bottom and two large insects above the soda lime (Figure 6-7). Be sure that the test tube is dry *before* intro-

TIME	SYRINGE READING	O₂ CONS.	ACTIVITY none → ext.

TABLE 1.

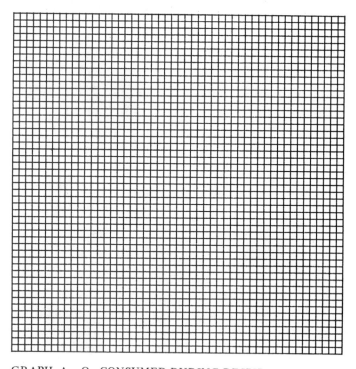

GRAPH A. O_2 CONSUMED DURING RESPIRATION.

ducing the animals, and make certain that the animals cannot come into contact with the soda lime.

2. Assemble a control tube with the same quantity of soda lime; the perforated container is not necessary. Fill with paper the same space as that taken up by the insects in the experimental tube.

3. Place a drop of fluid into the end of the capillary tube.

4. Allow the vent tubes to equilibrate to the air for five minutes.

5. Close the vent tubes with pinch clamps.

6. Read the changes at two-minute intervals for thirty minutes. Record the data in Table 1. As the experiment proceeds, note the activity of the animals (from "none" to "extreme"), and record in the table each time you take an "oxygen consumed" measurement.

7. Using your data, plot on Graph A the oxygen consumed against time. Be sure to label the axes properly. What was the total oxygen consumption? What is the average oxygen consumption over a two-minute period? With an estimate of the weight of the animals, given to you by the instructor, calculate the volume of oxygen consumed per gram of tissue per hour.

QUESTIONS FOR FURTHER THOUGHT

1. What are the functions of the rib cage and diaphragm in mammalian breathing?

2. From your data on insect respiration, what can you conclude about the relationship between the animal's activity and the amount of oxygen consumed?

3. What factors regulate the rate of gas exchange in a mammal?

4. What role does atmospheric nitrogen play in gas exchange?

TOPIC 7 — GAS EXCHANGE AND TRANSPORT IN PLANTS

Biological Science, 4th ed.
Chaps. 11 and 12, pp. 272–280, 295–317
Elements of Biological Science, 3rd ed.
Chaps. 10 and 11, pp. 171–176, 185–200

The cell's requirement for a supply of oxygen and nutrients, and for a means of disposing of wastes, makes necessary some sort of mechanism whereby these materials may be (1) exchanged with the environment and (2) transported from the site of exchange to other parts of the organism. In this laboratory, you will study the mechanisms of gas exchange in green plants as well as the structural adaptations of land plants for internal transport.

GAS EXCHANGE

While individual isolated cells of a plant are unable to regulate water loss, the leaf as a whole is adapted by **transpiration** to maintain water loss at an appropriate level, while providing as much area as possible for gas exchange and for the absorption of light. Special groups of cells in the epidermis—the **guard cells**, which bound the **stomata**—regulate the degree to which the interior of the leaf is open to exchange water and other materials with the atmosphere. By studying the photomicrograph (Figure 7-1), review the relationship of a stoma to the leaf interior and re-examine the structure of guard cells.

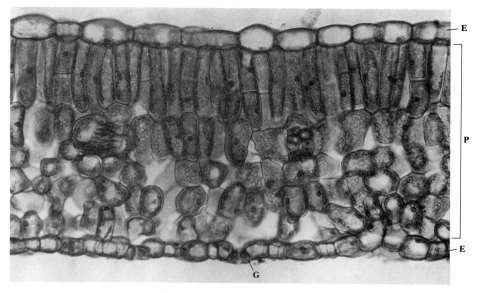

Fig. 7-1. Cross section of privet leaf. *(Photograph: Courtesy Thomas Eisner, Cornell University.)*

Adapatations to Special Habitats The privet leaf, which you have examined before, is a **mesophyte,** a plant that is adapted to a moderate type of habitat, neither very wet nor very dry. Extreme environmental conditions have sometimes engendered adaptations for gas exchange quite different from those of the privet leaf.

Nerium *leaf*   Examine a cross section of oleander (*Nerium*) leaf. Oleander is a **xerophyte,** a plant adapted for life in a very dry habitat. Note the prominent cuticle, thick epidermis, and stomata located in pits lined with epidermal hairs. Explain the functions of these adaptations, and compare this leaf with that of a mesophyte.

Potamogeton *leaf*   Study a cross section of the pondweed (*Potamogeton*) leaf. *Potamogeton* is a **hydrophyte,** a plant adapted for life in a very wet habitat. Note the stomata only in the upper epidermis, very thin cuticle, chloroplasts in the epidermis, and very large air chambers in the spongy mesophyll. Which of these adaptations would you consider to be evidence that this is a floating plant?

Unknown *leaf*   Your instructor will supply a leaf cross section from a plant of unknown or unspecified habitat. Study its structure carefully, determine the type of habitat to which it is adapted, and be able to give reasons for your choice.

Lenticels   Gas exchange in stems is accomplished partly through exchanges with the vascular tissues and partly by means of **lenticels,** loose groups of cells in the otherwise impervious epidermis or corky layer of a woody stem. Examine lenticels with a dissecting microscope. Figure 7-2A shows them on a maple (*Acer*) twig and Figure 7-2B in cross section of an elderberry (*Sambucus*) stem.

A

B

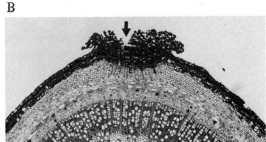

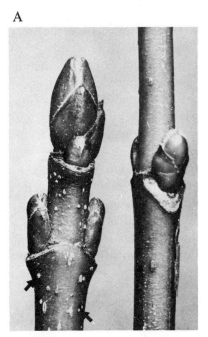

Fig. 7-2. Lenticels, indicated by arrows. A: On maple twig; B: in cross section of elderberry. *(Courtesy Carolina Biological Supply Company.)*

ABSORPTION OF RAW MATERIALS IN ROOTS

Almost all the water and minerals required by a plant are absorbed by the roots, while carbon dioxide, the key raw material in carbohydrate synthesis, is absorbed principally by the leaves.

Root systems

Examine demonstrations of several different types of root systems. Each demonstration will be accompanied by an explanatory card.

Root hairs

Obtain a radish seedling from the supply desk, or examine a seedling provided in a petri dish at your table. These seedlings have been grown in such a way that the roots are free from soil particles. If you examine the seedling closely, you will see that the part of the root closest to the tip is covered with numerous slender and delicate threadlike structures, the **root hairs.** These hairs, which are extensions of epidermal cells, greatly increase the absorptive area of a root. Water and dissolved inorganic salts of such elements as potassium, nitrogen, phosphorus, sulfur, calcium, magnesium, and iron all enter through the root hairs. Dissolved gases, such as the oxygen required for the respiration of root cells, also enter through the root hairs. Make a wet mount of the seedling, and observe it through a compound microscope. How thick is the cell wall? What are the methods by which water and dissolved minerals are absorbed by the roots?

Cellular structure of the root

Examine a slide of a cross section of buttercup (*Ranunculas*) root. Locate on the slide and label in Figure 7-3 the following regions:

1. The **epidermis,** which forms the continuous protective outer layer of the plant body. The cells are generally not large, and their walls, particularly the outer ones, are thick.
2. The **cortex,** which is inside the epidermis but outside the vascular region. It forms most of the bulk of the young root. The cells of the cortex are **parenchyma,** a tissue type composed of relatively unspecialized cells that are usually spherical in shape. They frequently have thin cell walls, and there are often numerous intercellular spaces between the cells. Note the many stored starch grains in the parenchyma cells.
3. The **endodermis,** which is a single layer of cells just inside the cortex. Endodermal cells have thickened cell walls and form the outermost portion of the **stele,** or vascular cylinder. What is the function of the endodermis?
4. The **pericycle,** which is a layer of thin-walled cells just inside the endodermis. These cells retain the ability to divide and give rise to branch roots.
5. The **phloem** tissue, which lies inside the pericycle as small, compact cells between the arms of the much larger cells in the middle. Phloem cells help to conduct food materials in the plant.
6. The **xylem vessels,** which are the large, thick-walled cells in the middle, whose arms prevent the phloem from making a continuous ring. They aid in the upward conduction of water and minerals, and provide support. (We shall examine both phloem and xylem tissues in more detail in a later laboratory.)

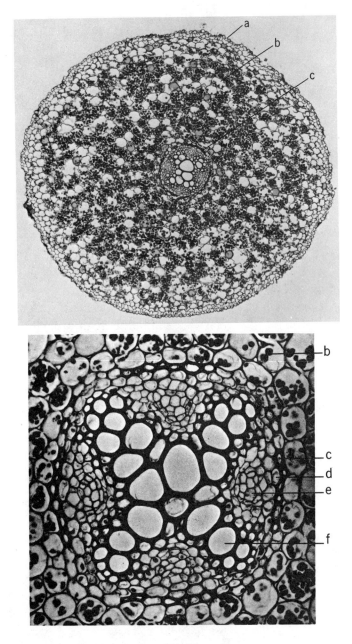

Fig. 7-3. Cross section of buttercup (*Ranunculus*) root. *(Top: Courtesy Thomas Eisner, Cornell University; bottom: Courtesy General Biological Supply House, Inc., Chicago.)*

VASCULAR TISSUE OF THE STEM

Many smaller organisms and some bigger ones have no special transport systems because diffusion and osmosis alone are sufficient for movement of substances across relatively short distances. In the case of larger animals and most large terrestrial plants, however, an efficient transport system involving the mass flow of materials is necessary. Local molecular traffic may still involve diffusion, osmosis, and active transport, but major distances are usually bridged by the movement of fluids through a more or less well-defined system of internal channels. In this section, we will consider the structural basis of transport in vascular plants.

The movement of fluids in plants occurs, for the most part, in two distinct types of **vascular tissue**. One of these, the **xylem**, usually carries water and dissolved inorganic materials upward from the roots to other parts of the plant, especially the leaves. There they are utilized in photosynthesis and other metabolic activities. The other vascular tissue is **phloem**, and it transports nutrients, particularly the organic products of photosynthesis, from one part of the plant to another. In addition to these predominantly vertical transporting systems, groups of undifferentiated parenchyma cells, called **rays**, serve in the larger plants for lateral transport across both xylem and phloem.

Xylem and phloem are complex tissues. Both consist of several different types of cells, some of which are alive and some of which are dead when functionally mature.

Corn stem

Before you study vascular systems in detail, it would help to examine a relatively simple plant in order to gain some idea of how the vascular tissues are related to the plant as a whole and to see what other sorts of cells we shall encounter in examining the vascular components of other plants.

Study a cross section of corn (*Zea*) stem under the dissecting microscope and using low and high powers of the compound microscope. Locate and label in the accompanying photomicrographs (Figure 7-4):

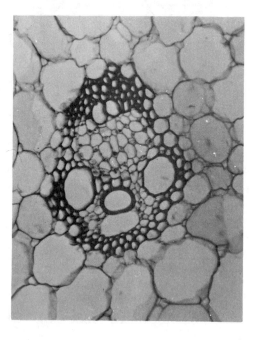
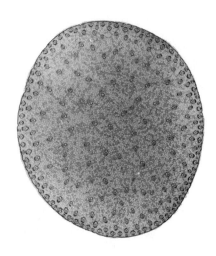

Fig. 7-4. Monocot (corn) stem cross section. *(Courtesy Thomas Eisner, Cornell University.)*

1. **Vascular bundles**, scattered throughout the stem but concentrated near its edge.

2. Thick-walled **sclerenchyma** cells surrounding each vascular bundle.

3. Large, open, thick-walled **xylem vessels**, usually two or three in number, toward the center of the stem.

4. **Phloem tissue**, located toward the periphery of the stem, containing thin-walled cells of two sizes:
  a. **sieve tube cells**, which function in transport of organic substances.
  b. **companion cells**, smaller cells adjacent to the sieve tube cells.

Scattering of the vascular bundles throughout the stem is characteristic of the **monocotyledons** (a subclass of the flowering plants, angiosperms; see Appendix 2 or your text), of which corn is a representative. Note how the bundles are oriented, with phloem always to the outside and xylem always to the inside. The tissues always differentiate in this manner, but the reason is unclear. Can you think of any explanation for this arrangement?

The sclerenchyma cells, thick-walled and usually stained a deep red, are supportive elements. Can you find sclerenchyma in the stem outside the vascular bundles? Where in the vascular bundles is this cell type most heavily localized?

Compare the thickness of wall and size of lumen (interior space) of sclerenchyma cells and xylem vessels.

The bulk of the corn stem, and not a few cells in the vascular bundles themselves, consists of parenchyma—thin-walled, undifferentiated cells without secondary thickenings.

Before considering the other types of arrangement of vascular tissues, we should examine briefly the structure of the transporting elements in phloem and xylem.

The Structure of Phloem and Xylem The transporting cell type in phloem tissue is the **sieve tube cell**, an elongate living cell without a nucleus and with many perforations in its end wall (Figure 7-5). Closely associated with the sieve tubes are smaller, elongate **companion cells**, which do have nuclei and probably exercise some metabolic control over the functioning of the sieve tube cells.

The transporting cells in the xylem are of two main types, **tracheids** and **vessels**, but within each of these groups there is wide variation. Both types

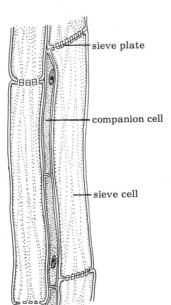

sieve plate

companion cell

sieve cell

Fig. 7-5. Longitudinal section of sieve tubes and companion cells.

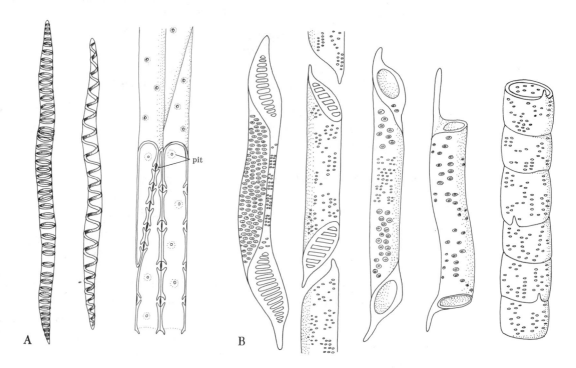

Fig. 7-6. Xylem. A: Tracheids of three different types; B: vessels of five different types, with the more primitive on the left and the more advanced on the right. In the more advanced cells, all traces of the end walls have disappeared. *(A: Modified from V. A. Greulach and J. E. Adams,* Plants: An Introduction to Modern Botany, *Wiley, 1962; B: Modified in part from A. J. Eames and L. H. MacDaniels,* An Introduction to Plant Anatomy, *McGraw-Hill, 1947.)*

are dead at maturity and have thickened secondary walls and a more or less cylindrical shape (Figure 7-6). Tracheids, spindle-shaped cells closed at the ends, have a smaller lumen than vessels. Vessels are usually wider in diameter than tracheids and tend to be heavily perforated at the ends or even open entirely. Tracheids often have **pits** in the walls and abut against one another more or less diagonally (Figure 7-6A). Vessels may have perforated walls, but they abut end to end, forming columns (Figure 7-6B).

Phloem and xylem cells   Examine a longitudinal section of a pumpkin (*Cucurbita*) stem. Locate the sieve tube cells and the companion cells. Draw and label the sieve tube cells and the companion cells. How do they compare to Figure 7-5? Locate the xylem cells, noting the arrangement of these cells. Draw and label the xylem cells. Do they abut one another end to end or diagonally? Identify as many other cell types as you can.

Macerated wood   Examine a slide of macerated wood (wood treated so as to dissociate the cells). Compare a conifer like pine (*Pinus*) with a hardwood like basswood or ash (*Tilia* or *Fraxinus*). Which cells are common in the conifer? In the hardwood? Locate as many different types of vessels and tracheids as time allows. If there is enough time, your instructor may give you specimens and directions for macerating your own wood specimens. In some of your slides, you may find tissues or cells not identifiable as tracheids or vessels. What are they?

Vascular Tissues in Other Stems You have already examined corn, a monocot, which has vascular bundles scattered throughout the stem. Some dicots (the other group of flowering plants; see Appendix 2) also have vascular bundles, but these are almost always organized into a ring at the periphery of the stem. Compare Figures 7-7 and 7-8, of the buttercup (*Ranunculus*) and alfalfa (*Medicago*), respectively. The significance of this arrangement will be apparent shortly. Most monocots and many dicots, like the two illustrated here, are known as **herbaceous** plants since their stems remain soft and succulent, in contrast to **woody** plants. (Do you notice any features of the buttercup stem [Figure 7-7] that suggest it is adapted to carry out photosynthesis?)

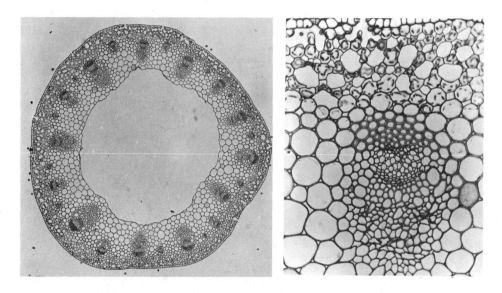

Fig. 7-7. Cross section of stem of buttercup, a herbaceous dicot. Left: Low power; right: high power. *(Courtesy George H. Conant, Triarch Incorporated, Ripon, Wisconsin.)*

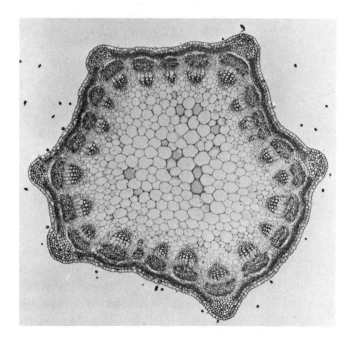

Fig. 7-8. Cross section of stem of alfalfa, a herbaceous dicot. *(Courtesy Thomas Eisner, Cornell University.)*

All the tissues in the corn stem, the buttercup, alfalfa, and a great many other herbaceous plants are known as **primary tissues**. By definition, primary tissues are those that differentiate directly from the **apical meristems**, the shoot tip and root tip of the plant. Primary tissue growth is sufficient for these plants since their relatively small stature does not require any substantial increase in girth.

Plants that do increase substantially in girth and often live through many growing seasons increase their diameter by reproduction of cells from a second type of meristem—the **vascular cambium**. This tissue can be seen in primitive form in the alfalfa stem; but usually the xylem and phloem are joined into a more complete ring, and a complete ring of cambium lies between them. As the cambium divides, new cells are added to the xylem on the inside and to the phloem on the outside. This **secondary tissue** is added to the stem or root in each growing season, and the organ's diameter and strength are thereby increased. Would the arrangement of vascular tissues in corn be adapted to the production of secondary tissues?

Woody dicot stem

Examine a cross section of basswood (*Tilia*), a woody dicot, with the dissecting microscope and under low power of the compound microscope. Label Figure 7-9 as you identify the parts of this stem's anatomy.

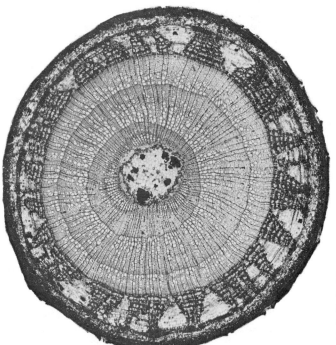

Fig. 7-9. Cross section of basswood stem at the end of the third year of growth. *(Courtesy Thomas Eisner, Cornell University.)*

1. From the center of the stem outward, the bulk of the stem consists of xylem tissue, arranged in **annual rings**. Each season's growth appears in a ring because of the difference in size of tracheids and vessels laid down at the end of one growing season and the beginning of the next. In a temperate climate, the beginning of the growing season is usually wet, and the xylem elements are large and are known as **spring wood**. The end of the growing season is frequently marked by drought, and the xylem elements formed in

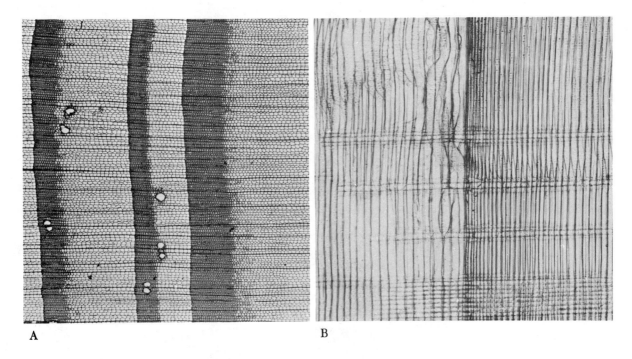

A   B

Fig. 7-10. Spring and summer woods. A: Cross section of larch; B: radial section of pine. Notice the fluctuation in tracheid size between spring and summer wood. *(Courtesy U.S. Forest Products Laboratory.)*

this part of the growing season are much smaller and are called **summer wood** (compare Figures 7-9 and 7-10).

How old is your stem? Can you think of conditions under which annual rings would not be formed?

2. The innermost xylem ring is primary tissue, formed by the apical meristem. Internal to the xylem, at the center of the stem, there is usually a small amount of parenchyma tissue called **pith.**

3. Each of the xylem rings outside the primary xylem was deposited by the vascular cambium. See if you can distinguish tracheids and vessels in the xylem.

4. Careful observation will reveal that the xylem is traversed by radially oriented groups of parenchyma cells. These are called **rays,** and their function is to move materials laterally through the stem.

5. The vascular cambium, a ring of tissue one cell thick, is best located under high power just outside the youngest ring of xylem.

6. Tissues just outside the cambium are functional phloem. (Except for the thick-walled sclerenchyma cells, cell types usually cannot be distinguished very easily.) The older phloem, farther out in the stem, is dead and does not function in transport. How does it function?

As one might expect, addition of cells on either side of the cambium causes the xylem and phloem to increase in diameter. As cells are added to the xylem on the outside of the cambium, the cambium expands to accommodate the increase in stem diameter. The older phloem, however, is fixed in size and cannot expand. As it is pushed outward by the advancing cambium, it finally breaks. Fill-in tissue is supplied by expansion of phloem parenchyma and expansion of the rays in the phloem, both of which are living cells

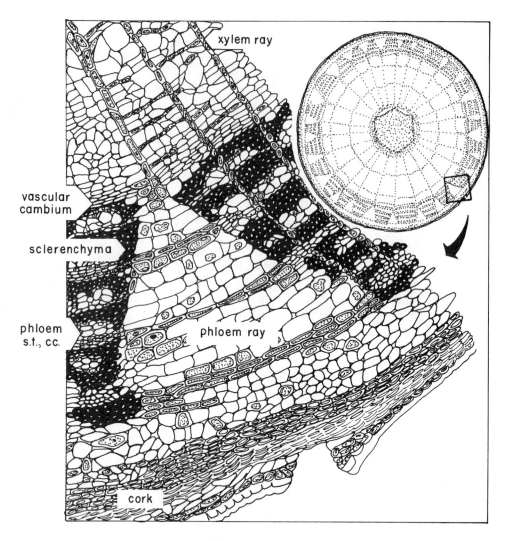

Fig. 7-11. Cross section through basswood (*Tilia*) phloem. The high-power drawing illustrates the progressive change in phloem from the vascular cambium to the outside of the stem.

capable of division. The breaking and fill-in is, however, evident in the appearance of the phloem: the triangular groups of cells in the phloem are partly old broken phloem, youngest to oldest, from the interior, and partly phloem ray and parenchyma expansion. Figure 7-11 shows a small portion of this region under high power.

7. As a stem matures, a number of changes also occur in its outer surface. The original **epidermis** is lost through weathering, and in the underlying **cortex** a new meristematic region, called the **cork cambium**, forms. This meristem produces cork cells, which become impregnated with a waxy substance and then die, thus forming an outer rough, water-repellent layer of the stem. As the cork breaks and sloughs off because of pressure from the vascular cambium and from weathering, the cork cambium is continually renewed a little deeper in the stem. Eventually, the original cortex weathers away completely; after this has occurred, new cork cambium forms from phloem parenchyma.

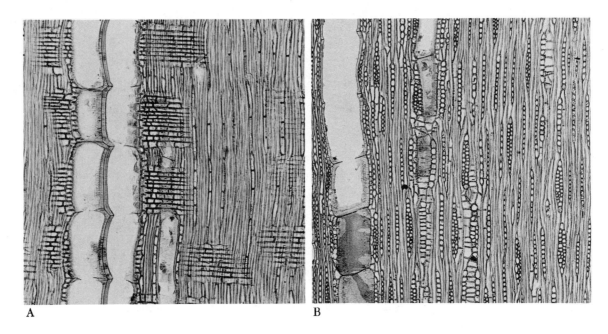

Fig. 7-12. Sections through persimmon wood. A: Radial; B: tangential. *(Courtesy U.S. Forest Products Laboratory.)*

8. Examine radial and tangial sections of *Tilia* stem. Locate and identify as many different vascular elements as you can. Label the photographs of these sections in Figure 7-12.

Remember that although we have used the stem as a basis for discussion, roots and leaves contain continuations of the transport system we have been studying, and that transport is no less important in those organs than it is in the stem.

EXPERIMENTAL STUDY OF TRANSPORT

Of the two transport systems in plants, xylem is far more amenable to experimental manipulation and is much better understood, both anatomically and functionally. Currently available evidence suggests that water and mineral transport results almost entirely from physical forces which natural selection has shaped to the plant's advantage. One of these physical forces is the remarkable cohesiveness of water molecules, which serves to keep water in the xylem in a continuous column, where other fluids would break under the negative pressure. The other force is simply cell-to-cell differences in osmotic pressure, which may tend to move water into or out of the xylem. One small source of osmotic pressure gradient is root pressure, the result of osmotic differences between the root-cell fluid and the soil. The major source, however, is transpiration, which you will study experimentally during this period.

Measurement of transpiration rate

1. Several months prior to the laboratory period in which this experiment is to be performed, you or your instructor should have prepared the experimental plant in the following manner. Take a cutting from a *Coleus*

plant, and put the stem through one hole of a two-hole stopper. Put the cut end of the stem into moist sand or potting soil in a flowerpot, and keep it moist and in sunlight until the cutting has regenerated a substantial root and shoot system.

2. Set up the apparatus as illustrated in Figure 7-13. Make certain that all connections are very snug. Remove the plant gently from its pot, immerse the root end in a large container of water, and swirl to wash the sand or soil from the roots. Fit the plant into a pint jar which you have previously filled with tap water. Press the stopper in to make a very tight fit. Be sure you have no air bubbles trapped in the jar. The water will be forced into the attached pipette. If more water is needed, use a small syringe to insert more water into the rubber coupling between the glass tubing and the pipette.

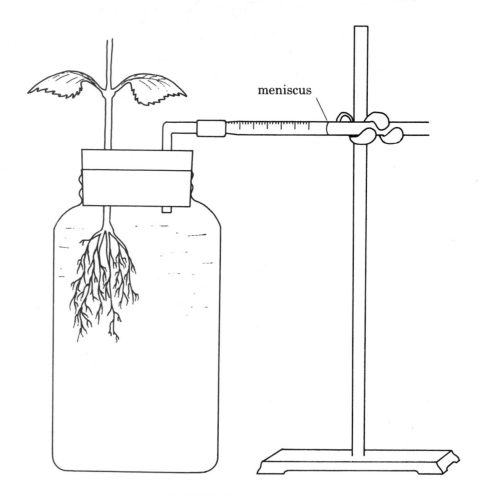

meniscus

Fig. 7-13. Potometer setup.

3. The rate of water loss can be calculated in milliliters per hour by measuring and recording changes in the position of the meniscus over known periods of time. Observe the approximate rate of movement for a while, and then decide on convenient intervals at which to take position and time readings. The precise volume and time interval you use are unimportant as long as you get sufficient data with which to construct a graph, and as long as the measurements you do take are accurate and consistent with one another.

After you have obtained several consecutive readings that agree well with one another, average them and record that figure in Table 1 as the normal transpiration rate.

4. With a strong light bulb, increase the light intensity near the plant, and continue to take readings for ten or fifteen minutes. Remove the light, and take "normal" readings again.

5. Record these data in the table as you go along. When you finish, plot them on Graph A, indicating the axes clearly and marking the curve to show where the light intensity was changed.

Additional experiments   Your instructor may specify, or you may wish to design, other experiments to measure the transpiration rate under different conditions. Some suggestions follow:

1. Compare transpiration rates under a wider range of different conditions of illumination intensity or under different qualities of illumination, or with "wind" (from a fan).

2. Compare the transpiration rates of plants that have had the stomata blocked (by coating one or both leaf surfaces with Vaseline) with plants that have not had the stomata so blocked.

3. Remove part of the root system from your original plant, and compare the plant's transpiration rate under these conditions with what you recorded when the root system was intact.

Rate of fluid movement   If there is time, you may wish to try to estimate the rate of fluid movement
in the xylem   in the xylem. At the supply table you will find a beaker containing celery stalks that have been immersed in a solution of 0.5% aqueous basic fuchsin for one hour. (Basic fuchsin is a powerful red dye. *Handle with care!*) Place one of the stalks on a folded paper towel, and, slicing it as thin as you can through the stem several inches from the bottom, make cross sections. (The sections should be transparent.) Mount the sections on clean slides in a small amount of water but without a cover slip. Examine the sections under the compound microscope. Measure the diameter of your low-power microscope field, and from this information make an estimate of the area occupied by the lumen of the vessels. With this knowledge, and your data on transpiration rate, you should be able to calculate the approximate rate of flow of water through the xylem.

QUESTIONS FOR FURTHER THOUGHT

1. What direct role does photosynthesis play in guard-cell operation?

2. Conifers have only tracheids, no vessels. Can you suggest some way in which tracheids might serve as a useful adaptation for a plant living in a very cold habitat?

3. What is the function of sclerenchyma in the vascular bundles of corn?

4. A corn stem shows no secondary growth. How does it manage to increase in girth?

PIPETTE READING	H₂O TRANSPIRED
AVERAGE RATE:	

NORMAL INTENSITY

PIPETTE READING	H₂O TRANSPIRED
AVERAGE RATE:	

HIGH INTENSITY

PIPETTE READING	H₂O TRANSPIRED
AVERAGE RATE:	

NORMAL INTENSITY

TABLE 1.

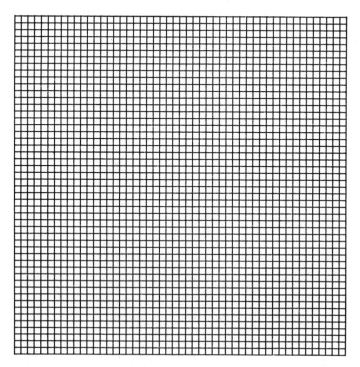

GRAPH A.

5. A nail is driven into a ten-year-old tree at 3 feet from the ground. The tree is 15 feet tall. When the tree is 20 feet tall, how far from the ground will the nail be? Why?

6. Discuss and compare the theories of translocation in the phloem, and be able to give evidence for each.

7. What happens to the phloem of a woody stem as the stem ages?

8. Discuss the leaf structures that help to adapt a xerophyte like oleander (*Nerium*) to its environment. Compare these adaptations to those of hydrophytic and mesophytic plants.

9. Describe the manner in which the age of a woody twig can be determined from external and internal observations.

10. Discuss the processes involved in the uptake of water and minerals by a plant root.

11. Compare and contrast secondary xylem and secondary phloem with regard to meristematic origin, location in the plant body, cellular characteristics, probable method of movement of materials, and general function in the life of the plant.

TOPIC 8 TRANSPORT IN ANIMALS

Biological Science, 4th ed.
Chap. 13, pp. 318–351
Elements of Biological Science, 3rd ed.
Chap. 12, pp. 201–219

Association and cooperation of diversely specialized cells make possible complex organisms like elephants and earthworms and vascular plants. The integrity of any organism depends upon mechanisms for functionally knitting together its widely separated parts, thereby promoting smooth and efficient operation of the whole system. We may properly regard any fluid transport mechanism, regardless of its structural basis, as an integrating mechanism. It not only carries raw materials and by-products of metabolism between sites of exchange with the environment and other parts of the body, but also serves as the medium in which a variety of chemical messages—hormones—are rapidly circulated throughout the organism.

You have already examined some adaptations for fluid transport in the vascular plants. In this unit, you will consider the anatomy of the heart and the circulatory system in a mammal. The structure and chemistry of blood will also be examined in this week's laboratory. In the next unit, you will have an opportunity to study capillary circulation and the control of heartbeat in a frog.

THE MAMMALIAN HEART AND GREAT VESSELS

Turn your attention again to the thorax of the rat that you previously opened. Locate the thymus gland, a lighter-colored tissue mass that lies anterior to the heart. A special class of white blood cells, the T lymphocytes, develop in the thymus gland and are responsible for cell-mediated immune responses.

Remove the thymus and any fat that may interfere with your observation of the heart and the vessels associated with it. If you have not already done so, carefully remove the pericardium, the thin membrane covering the heart.

Examining the heart
    Heart   The right and left **ventricles** compose the major portion of the heart visible from the ventral side. Anterior to the ventricles, you can see the darker-colored **atria** (singular, "atrium"). The atria are more readily visible if the heart is pushed to one side and examined dorsally (Figure 8-1).

The heart may be sectioned frontally, separating ventral from dorsal side, and examined under a dissecting microscope for the internal anatomy of the chambers and valves (Figure 8-2). Make incisions with a scalpel or razor blade, and cut the chambers open far enough to see the internal structure clearly. If a beef, calf, or sheep heart is available, it will be easier to examine the internal anatomy. Carefully examine the interior of all four chambers, locating the following valves:

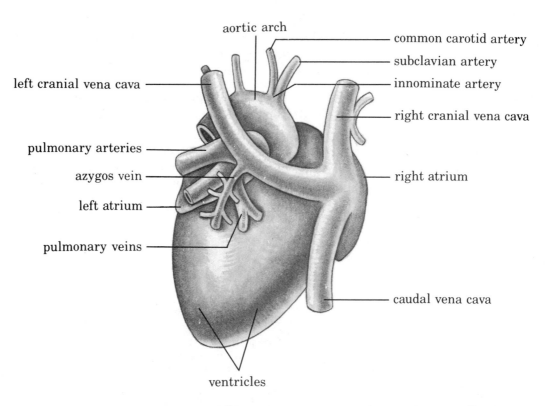

aortic arch

common carotid artery

subclavian artery

left cranial vena cava

innominate artery

right cranial vena cava

pulmonary arteries

azygos vein

right atrium

left atrium

pulmonary veins

caudal vena cava

ventricles

Fig. 8-1. Rat heart—dorsal view. *(Modified from Chiasson, Carolina Biological Supply Company.)*

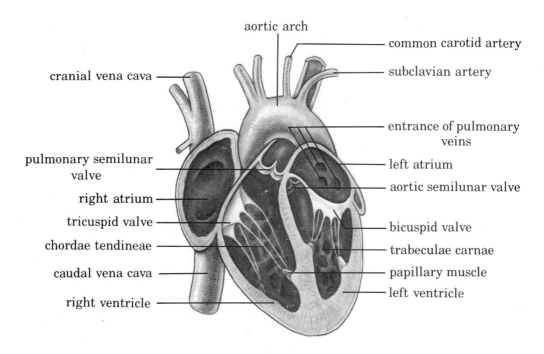

aortic arch

common carotid artery

subclavian artery

cranial vena cava

entrance of pulmonary veins

pulmonary semilunar valve

left atrium

aortic semilunar valve

right atrium

tricuspid valve

bicuspid valve

chordae tendineae

trabeculae carnae

caudal vena cava

papillary muscle

right ventricle

left ventricle

Fig. 8-2. Rat heart—sagittal view. *(Modified from Chiasson, Carolina Biological Supply Company.)*

1. **Tricuspid valve**, between the right atrium and right ventricle.
2. **Bicuspid** or **mitral valve**, between the left atrium and left ventricle.
3. **Pulmonary semilunar valve**, in the base of the pulmonary trunk.
4. **Aortic semilunar valve**, in the base of the aorta.

How do the tricuspid and bicuspid valves differ? Notice the muscular bands attaching the valve flaps (cusps) to the ventricle walls. What is the function of these bands?

Review the function of each of the valves and the passage taken by blood through the heart of a mammal. Label Figure 8-3 with arrows showing the direction of blood flow. How do the hearts of other vertebrates differ from that of mammals?

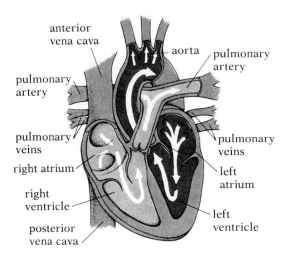

Fig. 8-3. Blood flow in the human heart.

Sometime during the period, if materials are available, your instructor will use the fresh beef heart to demonstrate the action of the heart valves, coronary circulation, and a simulated coronary thrombosis.

Arteries   Locate the following vessels:

Examining the great blood vessels

1. Pulmonary trunk, leaving the ventral side of the heart from the top of the right ventricle, then passing forward diagonally before branching into right and left **pulmonary arteries**, which carry blood to their respective lungs. Trace one of the pulmonary arteries as far as you can into the lung.

2. **Aorta**, arising from the anterior end of the left ventricle, just dorsal to the origin of the pulmonary trunk, then passing a short distance forward and bending to the *animal's* left as the **aortic arch**.

Veins   In your specimen, identify the following veins:

1. Right and left **anterior vena cavae**, bringing blood from the upper parts of the body and entering the right atrium at its anterior end (Figure 8-1). In humans, a single anterior vena cava drains blood from the upper part of the body.

2. **Posterior vena cava,** bringing blood from the lower parts of the body and entering the right atrium at its posterior end.

3. **Pulmonary veins,** returning blood from the lungs to the left atrium.

Note the difference in appearance and in texture between veins and arteries, and compare their structure with that of a capillary in Figure 8-4.

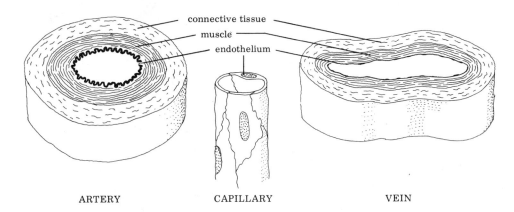

ARTERY   CAPILLARY   VEIN

Fig. 8-4. A comparison of the walls of artery, vein, and capillary. Arteries and veins have the same three layers in their walls; but the walls of veins are much less rigid, and they readily change shape when muscles press against them. Capillaries have walls composed only of a thin endothelium.

SYSTEMIC CIRCULATION IN THE RAT

The systemic circulation serves all of the body except for the lungs. It consists of arteries, which branch from the aorta all along its length, veins, which return blood from all over the body to the right atrium, and capillaries between the arteries and veins. Although you will examine a few of the major arteries and veins to get an idea of the pattern of systemic circulation in a mammal, do not forget that the exchanges between the blood and tissues are the principal function of the circulatory system, and that such exchanges occur across the walls of the capillaries. The major vessels and the heart function only to deliver blood to the capillary beds.

Trace each of the vessels as far as you can.

Arteries   (Figure 8-5)

Examining arterial circulation

1. The first branches off the aorta are two **coronary arteries,** which leave the aorta as they cross dorsal to the pulmonary trunk. One of them is visible on the ventral surface of the heart. These arteries supply capillary beds in the heart wall, since its needs are not met by blood passing through the chambers of the heart. The next branch from the aorta, the **innominate artery,** divides promptly into the **right common carotid artery,** serving the head and neck, and into the **right subclavian artery,** serving the right forelimb. The **left common carotid artery** branches directly from the aorta and also serves the head and neck. The **left subclavian artery,** branching to the left forelimb, also arises directly from the aortic arch.

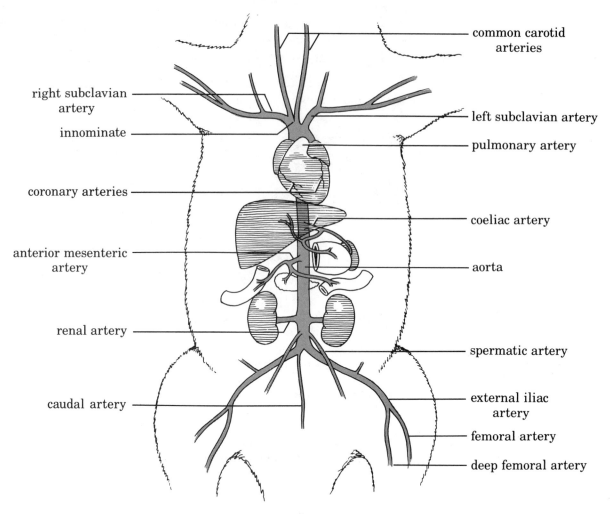

right subclavian artery

innominate

coronary arteries

anterior mesenteric artery

renal artery

caudal artery

common carotid arteries

left subclavian artery

pulmonary artery

coeliac artery

aorta

spermatic artery

external iliac artery

femoral artery

deep femoral artery

Fig. 8-5. Rat arterial system.

2. Follow the aorta as it descends along the middorsal line. It gives off several major branches:

    a. The first major branch from the aorta in the abdominal cavity is the **coeliac artery,** a short vessel supplying the stomach, duodenum, liver, and spleen. Trace branches of the coeliac into the respective organs as far as you can.

    b. Another unpaired branch of the aorta, arising a short distance posterior to the coeliac artery, is the **anterior mesenteric artery,** which branches to the mesenteries and small intestine.

    c. There is a pair of **renal arteries,** one supplying each kidney.

Near the posterior end of the aorta, paired arteries branch to the organs in the lower abdominal cavity and to the hind legs.

Veins   (Figure 8-6)

Examining venal circulation

1. Both the right and left **anterior vena cavae** return blood to the right atrium from the head, neck, thoracic limbs, and a substantial part of the

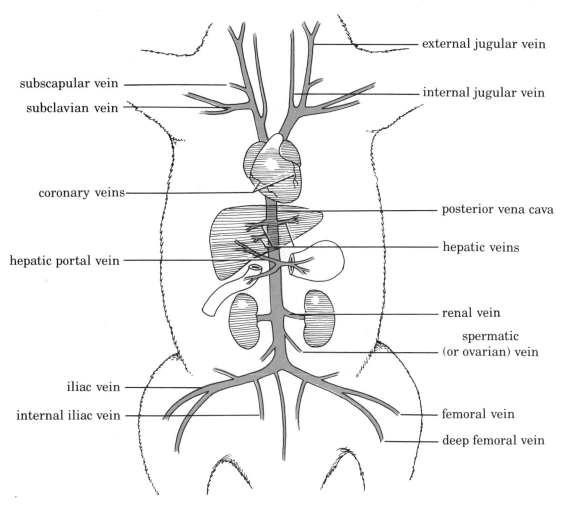

external jugular vein

subscapular vein

internal jugular vein

subclavian vein

coronary veins

posterior vena cava

hepatic veins

hepatic portal vein

renal vein

spermatic (or ovarian) vein

iliac vein

internal iliac vein

femoral vein

deep femoral vein

Fig. 8-6. Rat venous system.

thoracic wall. The more prominent veins that drain into each of the anterior vena cavae are the **subclavian** and the **jugular veins.**

2. The **posterior vena cava** is a large vessel that returns blood from the posterior part of the body to the right atrium. It receives no branches in the thoracic cavity, but it does receive two or three small veins from the diaphragm. Several large **hepatic veins**, draining capillary beds in the liver, enter the posterior vena cava just before it enters the liver. Right and left **renal veins** drain the corresponding kidneys and lie adjacent to the renal arteries. A pair of veins draining the hind legs unite in the posterior part of the peritoneal cavity to give rise to the posterior vena cava. These veins form from a number of tributaries that come from the hind limbs and other posterior parts of the body.

Hepatic Portal System Veins normally carry blood from organs to the vena cava and back to the heart. In certain pathways, the blood passes from the arteries to capillaries to veins to another network of capillaries and veins

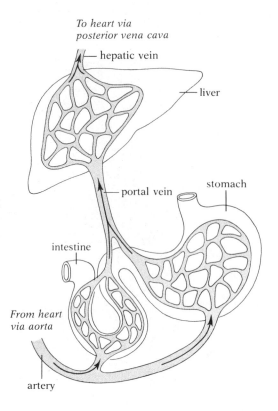

To heart via posterior vena cava

— hepatic vein

— liver

portal vein

stomach

intestine

From heart via aorta

artery

Fig. 8-7. The hepatic portal circulation. Blood from the capillaries of the digestive tract and stomach is carried by the portal vein to a second bed of capillaries in the liver. It then flows via the hepatic vein into the posterior vena cava, which takes it to the heart.

and finally returns to the heart. This type of system is called a **portal system** (Figure 8-7) and is defined as a group of veins that leads from one capillary network to another. In the rat, veins leading from the stomach, pancreas, spleen, and intestinal capillaries to veins in the liver constitute the **hepatic portal system.** The significance of this circulatory arrangement is apparent if one considers the important role of the liver in processing and storing materials absorbed from the intestine.

Veins from the stomach, pancreas, and spleen converge to form a vein that unites with the vein draining the intestine, forming the **hepatic portal vein.** The hepatic portal vein runs dorsal to the pylorus and enters the liver. Blood leaves the liver via the **hepatic veins,** enters the vena cava, and returns to the heart.

BLOOD

Examining blood

The first of the following experiments and observations should be carried out by individual students. The others may be done by individuals or as demonstrations, as time permits and the instructor directs.

Microscopic Examination of Blood   Clean a microscope slide with 70% alcohol. With a *sterile* lancet (*open one yourself*–don't use one already opened or employed by someone else!), obtain a drop of your own blood (alternatively, beef blood may be used if available), and let it fall on one end

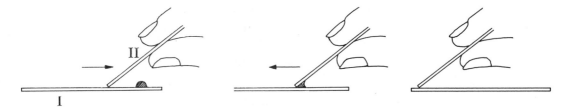

Fig. 8-8. Method for making a blood smear.

of the slide. Make the puncture in the end of your index finger, and squeeze the finger toward the end if enough blood to make a large drop does not promptly appear. Discard the lancet into the wastebasket or as directed by your instructor. With a wax pencil, make some identifying mark on your slide so you will know which is the side with the specimen on it.

As shown in Figure 8-8, take a second slide, and place it at an angle on the first slide. Move the second slide toward the drop of blood until the slide touches the drop; blood will flow along the edge of slide II. Then move slide II back along slide I, drawing the blood out into a thin film or *smear* across slide I. Allow slide I to dry, place it in a petri dish, and add enough Wright's stain to cover the blood smear. Count the number of drops of stain you add. After one minute, add an equal number of drops of distilled water. Leave the stain and water on for three more minutes, and then rinse the slide off by dipping it in a beaker of clean tap water. Allow the stained slide to dry thoroughly, smear side up, and then examine it under the microscope.

Most of the cells you see will be **erythrocytes,** the red blood cells. The **leukocytes,** white cells, are rarer, scattered among the erythrocytes. The platelets are destroyed in making the slide and will not be visible. Try to identify as many different kinds of leukocytes as you can, and draw them. Your instructor will be able to provide references that will enable you to identify these cells by name.

The Composition and Clotting of Blood   You will be provided with beef blood to which sodium oxalate has been added. Sodium oxalate binds calcium ions in the blood and thus prevents clotting, since clotting requires calcium ions. The oxalate does not otherwise interfere with these experiments.

1. Place some oxalated blood in a test tube, and allow it to stand undisturbed for a few minutes. Alternatively, it may be put into a centrifuge tube and centrifuged. Notice that a red mass sinks to the bottom of the tube. What is in this mass?

2. Remix the separated blood or obtain some fresh oxalated blood, add calcium-chloride solution, and watch the reaction. Addition of an excess of calcium ions (more than can be bound by the available sodium oxalate) will bring about prompt clotting. After clotting has occurred, centrifuge again, or allow the blood to stand. How does the remaining **serum** differ from plasma?

The Effect of Oxygen and Carbon Dioxide   Bubble air into 15 ml of oxalated blood in a test tube, and observe the color that develops. Then bubble carbon dioxide into the blood, note the color change, and again bubble air into the tube. How do you explain the changes in color? What reactions were taking place? How are oxygen and carbon dioxide normally carried in the blood, and under what conditions are they released?

The Buffering Action of Plasma   Maintenance of proper pH is of utmost importance to an organism, since even small shifts in the hydrogen-ion concentration can cause death. A stable pH is maintained by **buffers,** substances that tend to prevent major shifts in pH by binding or releasing hydrogen ions as the solution deviates from the "standard" for that buffering solution. Review the activity of buffers in your text, and carry out the following experiment, which demonstrates the action of buffers in the blood plasma.

1. In one beaker, place 3 ml of plasma obtained by centrifuging oxalated beef blood. In a second beaker, place 3 ml of distilled water. Make sure the beakers are properly labeled.

2. Use Hydrion test paper to determine the approximate pH in each beaker, using a different piece of paper for each test. If there is a difference in pH between the two beakers, adjust the pH of the distilled water by adding drop by drop 0.1 N hydrochloric acid (to reduce pH) or 0.1 N sodium hydroxide (to raise pH), testing after each addition, until the water has the same pH as the plasma.

3. When the plasma and the water have the same pH, add 0.1 N hydrochloric acid to the plasma drop by drop, swirling and testing after each addition, until the pH drops two units. Record the number of drops required for this change. Repeat this procedure with the water, again counting the number of drops required for a two-unit alteration in pH. Which of the solutions requires more acid to alter its pH? Why?

Hemolysis   Maintenance of the proper osmotic pressure of the blood is important. What organs help control the osmotic pressure of the blood? The following experiment demonstrates the effects of a change in osmotic pressure on the erythrocytes.

Place 1 ml of oxalated blood into each of three test tubes, and number the tubes. Into tube 1, place 10 ml of isosmotic (0.9%) saline solution; into tube 2, place 10 ml of distilled water; and into tube 3, place 10 ml of 5% saline solution. Mix the contents of each tube by shaking the tube gently.

Do you notice any difference in the appearance of the tubes? Allow them to stand for five minutes, stir again, and make a wet mount of each. Examine the slides under the microscope, and record your observations in the form of sketches. Explain the results.

QUESTIONS FOR FURTHER THOUGHT

1. Trace a drop of blood in the rat from the right atrium to the small intestine by the shortest possible route.

2. What is the significance of the hepatic portal system?

3. The blood of a human being serves as a medium of internal transport for many substances, yet it is critically important to the organism that fluctuations in the content of this fluid be held to a minimum. Consider such materials as water, glucose, amino acids, oxygen, carbon dioxide, inorganic salts, and urea, and indicate:

    a. Where each enters the blood.

    b. Where each leaves the blood.

    c. Measures the body can utilize to raise the concentration of each.

    d. Measures the body can utilize to lower the concentration of each.

(Don't forget the roles of hormones in these regulatory functions.)

TOPIC **9**

Biological Science, 4th ed.
Chap. 13, pp. 318–351
Elements of Biological Science, 3rd ed.
Chap. 12, pp. 201–219

HEARTBEAT AND BLOOD FLOW

In this period you will have an opportunity to observe a living circulatory system and to demonstrate basic aspects of the control and activity of the heart. The main objectives of this exercise are:

1. To observe blood flow in the capillaries.
2. To study the heart in action and map out its activities.
3. To demonstrate the effect of temperature on the rate of heartbeat of an amphibian.
4. To study the nervous control of the heart and to test the effect of some chemical transmitter substances and other drugs on heartbeat.

BLOOD FLOW IN THE CAPILLARIES

The major function of the circulatory system is served by capillaries, which connect the arteries and veins. In the frog, there are three areas transparent enough to enable you to see blood flowing in the capillaries: the web of the foot, the mesentery, and the tongue. Of these, the web is most convenient and can be examined under the microscope without injury to the animal.

1. The frog may be anesthetized or wrapped in a towel, but it is more easily handled if the nervous system is inactivated by pithing. (That the heart continues to beat after pithing demonstrates that the basic rate of heartbeat is endogenous to the organ; the brain simply modifies the rate when appropriate.) If you wish to pith a frog, inform your instructor, and consult Appendix 3.

2. Place the frog belly down on a balsa frog board (Figure 9-1) so that the tip of the foot just reaches over the circular hole. Stretch a part of the web over the hole, pinning the web so that a portion of it is fixed taut. Place the board on the stage of a compound microscope, fastening the board to the stage with a stage clip, a rubber band, or masking tape. Study the web portion with transmitted light, keeping the web moist with frog Ringer's solution.

Observations within the capillary net

1. The **chromatophores** are black pigment cells in the skin. Some of them are irregularly branched; others are compact. The expansion and contraction of the pigment is under hormonal control. The hormone primarily responsible is **intermedin** and is secreted by the pituitary gland. The color of the frog is light when the pigment is contracted and dark when it is expanded.

Can you see both expanded and contracted chromatophores in the same field? Why is regulation of color important in frogs? Can you think of

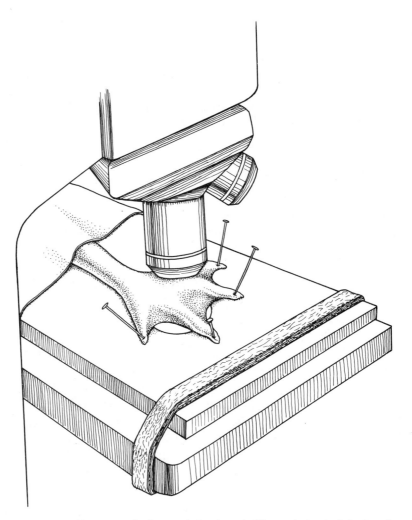

Fig. 9-1. Frog attached to a balsa board. The web is stretched and pinned over the hole; a rubber band holds the board to the microscope stage.

other animals that utilize color in this way? Why is regulation of color *not* important in humans?

2. There is a close network of blood vessels in the web, lying deeper than the pigmented layer.

 a. **Arteries** run mainly toward the free edge of the web, constantly diminishing in size as they break up into branches. The blood flows from larger to smaller branches, but remember that directions are reversed under the microscope.

 b. **Capillaries** are the very numerous tiny vessels, frequently branching.

 c. **Veins** are formed from the union of capillaries and increase in size as smaller veins unite with one another. Blood in veins flows from smaller to larger vessels, but remember about optical inversions.

 d. **Lymph vessels** are found in nearly all parts of the body as delicate, thin-walled, branching tubes. Lymph vessels may sometimes be seen in preparations. They originate in lymph capillaries, interwoven with, but without direct connections to, blood capillaries.

These vessels pick up fluids exuded from the blood in its passage through the tissues, and they eventually return the fluid to the circulatory system.

Blood flow is characterized by the presence of erythrocytes—red blood corpuscles. The flow is most rapid in the arteries, and slowest and most uniform in the veins. Trace the movements of the corpuscles through the capillary net.

HEARTBEAT

When one considers the large amount of blood that is pumped through the extensive system of vessels, it is easy to understand why a strong pumping device is necessary. You will now observe this pump in action, map the path taken by blood passing through the heart, and examine some factors that influence the rate of heartbeat.

The frog heart and its action

1. Remove the frog from the balsa board, place it ventral side up on a moist paper towel in a dissecting pan, and pin the forelimbs securely to the pan. Make a slit through the *skin only,* from the pelvis to the throat, along the mid-line. At each end, cut to the right and left, making two flaps. The abdominal vein and muscles will now be exposed. Following the cuts made in the skin but *avoiding* the ventral abdominal vein, cut the muscle of the abdominal wall. Anteriorly, keeping the scissors point up, cut through the pectoral girdle and up toward the throat.

Now observe the heart. Use cotton moistened in frog Ringer's solution to sponge away blood and to keep all exposed organs moist at all times.

2. With fine scissors and forceps, very carefully split open the pericardium—the sac covering the heart—to expose the heart (Figure 9-2). Note the following:

    a. Ventricle (a single, posterior, cone-shaped, thick-walled region).

    b. Arterial trunk (arising from the right side of the base of the ventricle and dividing anteriorly into two aortic arches).

    c. Atrium (a thin-walled, rounded chamber anterior to the ventricle and interiorly subdivided into two chambers, right and left).

    d. Sinus venosus (an elongated, thin-walled sac lying beneath the atria; it receives blood from the major veins and transmits it to the atria).

In bird and mammal hearts, the sinus venosus is not present as a separate chamber.

After examining the chambers of your frog's heart, label the dorsal and ventral views of Figure 9-3, indicating the sinus venosus, left atrium, right atrium, ventricle, arterial trunk, and aortic arches.

3. Watch the hearbeat carefully; it is a regular series of alternating contractions. The two atria contract together, the ventricle follows, and the arterial trunk contracts last. Gently raise the ventricle to see the sinus venosus; it will contract immediately before the atria.

The sequence of events in one heartbeat is as follows: The sinus venosus receives blood from the **vena cavae**, the major veins of the systemic circula-

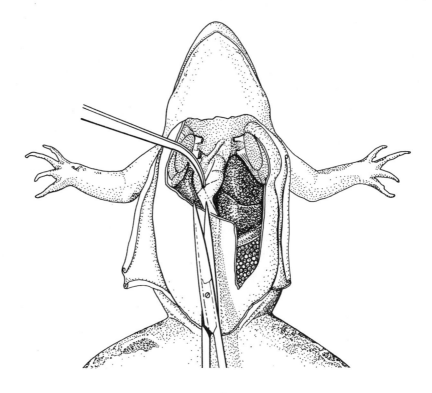

Fig. 9-2. Dissection just prior to opening the pericardial sac.

tion, and passes it on to the right atrium. The left atrium receives blood from the lungs (high in oxygen content) via the **pulmonary veins**. Both atria empty into the single ventricle, from which blood is forced by contraction of the ventricle wall. Because the blood only remains in the heart a short time, the two kinds of blood do not mix much. Venous blood on the right side of the ventricle is nearest the outlet and enters the arterial trunk first. There is partial separation of the streams within the arterial trunk, venous blood tending to go to the pulmonary arteries and oxygenated (pulmonary) blood tending to enter the systemic circulation. There is some mixing of the circulations and, thus, some difference in efficiency between this type of circulation and that of a mammal, in which there is complete separation of the pulmonary and systemic circuits. Can you suggest a reason why the incomplete separation in amphibians might be functionally advantageous to their way of life?

With arrows, indicate on the cutaway ventral view in Figure 9-3 the direction of blood flow through the heart.

When you have seen all the structures of the heart and have carefully observed its beat, review the other internal organs of the frog.

The Effect of Temperature on Rate of Heartbeat In vertebrates, a distinction is commonly drawn between cold-blooded animals (fish, amphibians, and reptiles) and warm-blooded animals (birds and mammals). Because such terms tell us nothing about how the temperature of the blood is regulated, we prefer to use the terms **ectothermic** when referring to animals whose tem-

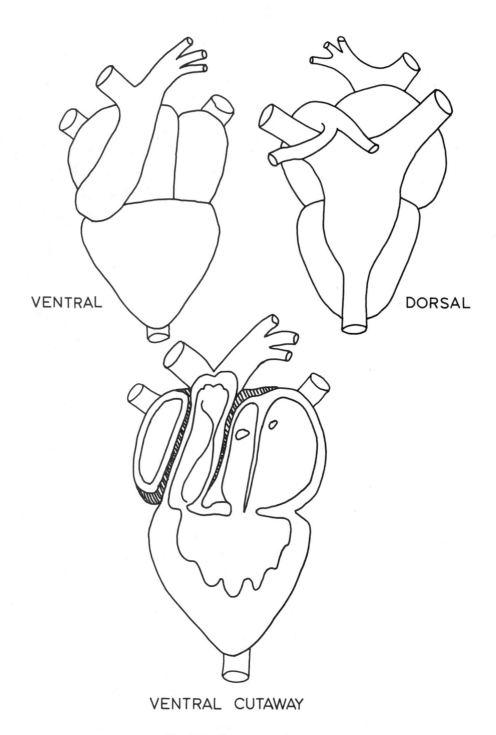

VENTRAL

DORSAL

VENTRAL CUTAWAY

Fig. 9-3. Frog heart—three views.

perature is regulated by the environment and **endothermic** when referring to animals with intrinsic temperature regulation.

1. Count the frog's heartbeats per minute at room temperature. Take the frog's temperature by placing the bulb of a laboratory thermometer carefully under the lobes of the liver.

2. Gently remove the frog on its towel, and cover the bottom of the dis-

secting pan with crushed ice. Make a frog-sized shallow depression in the ice, and place the frog, still on its towel, into the depression, ventral side up. Quickly insert a thermometer under the liver, bracing it with some object so that you do not have to hold it throughout the experiment and so that the weight of the thermometer does not interfere with normal heartbeat. Record the heartbeat rate and the temperature at frequent intervals until the temperature has dropped to 1°C or 2°C (or becomes stationary at 3°C or 4°C for several minutes), remove the thermometer, and empty the pan of ice. Reinsert the thermometer as before, and brace it. Record the heartbeat rate and the temperature as the frog's body temperature returns to room temperature.

In your report, plot the frog's heartbeat rate against body temperature, and discuss the results of the experiment with regard to what you already know about the frog's changeable metabolic rate. How might these circulatory responses to cold temperatures be of advantage to the animal?

Nervous Control of Heartbeat   Although the heart is capable of beating in the absence of nerves, the rate of heartbeat can be controlled by two sets of nerves that originate in two different parts of the central nervous system. One set is from the sympathetic nervous system, and the other from the parasympathetic. In time of danger or fear, the sympathetic system prepares the body for "fight or flight" by stimulating accelerator nerves that increase the heartbeat rate. The causative agent is a chemical substance (noradrenalin) released by the accelerator nerves.

The increase in heartbeat rate and in blood pressure would be harmful if prolonged. When the danger is over, the parasympathetic system slows down the heart. Vagus nerves of the parasympathetic system originate in the medulla of the brain and terminate in the sinus node, located on the sinus venosus.

The instructor will demonstrate the effects of vagus stimulation. One of the vagus nerves is ligatured (tied with thread) at a point near the brain and is cut away from the medulla. The nerve may then be lifted out of the body and stimulated electrically without stimulating any other organ. Ask the instructor what voltage he or she is using and how many impulses per second. What are your observations? Stimulation of the vagus causes nerve endings at the sinus node to release the chemical transmitter acetylcholine.

The Effect of Transmitter Substances   The effect of transmitter substances on heartbeat rate will be first observed on the intact heart.

1. After recording the normal rate of beat, place three or four drops of 5×10^{-5} M solution of acetylcholine directly on the heart. Record the rate of beat until a perceptible change occurs. If necessary, add more acetylcholine solution.

2. Rinse the heart thoroughly with Ringer's solution, using cotton to absorb the transmitter substance. Allow the beat to return to normal. This should occur within a few minutes.

3. Place three or four drops of 5×10^{-5} M adrenalin solution. The effects of adrenalin are like those of noradrenalin. Record the heartbeat rate, and

carefully watch the nature of the beating. Does adrenalin affect the heart in any way other than rate of beat?

4. Rinse the heart with Ringer's solution, and wait five minutes for its beat to return to normal. The effect of adrenalin wears off more slowly than that of acetylcholine.

5. Place on the heart several drops of 5×10^{-4} M solution of atropine in Ringer's solution. After thirty to sixty seconds, record the rate of beat. Add two to four drops of acetylcholine, and record again. Describe the results.

Now remove the heart by cutting as widely as possible around the sinus venosus, veins, and arteries. Place the excised heart in a depression slide of Ringer's solution, and test the effect of the transmitter substances on the excised heart.

1. After recording the normal rate of beat, add three to four drops of acetylcholine. Record the beat.

2. Rinse the heart thoroughly by dipping it in fresh Ringer's solution. Allow the heart to return to normal.

3. Transfer the heart to a second depression slide to which you have added three or four drops of the adrenalin solution. Record your results.

4. Return the heart to the acetylcholine solution, and note the rate.

5. Rinse the heart in Ringer's solution, place the heart on a third slide with the atropine solution, and note the results.

Compare your results on the excised heart with your results on the intact heart.

(*Note:* If at any time during the experiments the heart fails, try to restore it with gentle massage. If this does not work, attempt restoration by "shock treatment"—6V at 10 impulses per second from your instructor's stimulator should be enough.)

Acetylcholine exerts its effect principally in the sinus venosus area. Adrenalin affects other cardiac cells as well as those in the sinus venosus area. As you have seen, adrenalin also affects the amplitude (intensity) of contraction. How might you record the amplitude of heartbeat? Atropine appears to inhibit the action of acetylcholine in the manner of an enzyme inhibitor by combining with the acetylcholine receptor site on the post-synaptic membrane in the sinus venosus.

QUESTIONS FOR FURTHER THOUGHT

1. What is the function and operation of the lymphatic system?
2. Explain the differences between blood flow in the veins and in the arteries.
3. Explain the chemical changes occurring at the sinus venosus after stimulation of the frog's vagus nerve.
4. Why does the heart continue to beat after it is removed?
5. On the basis of your experience with the frog heart, suggest some probable difficulties which might be encountered in human heart transplants.

TOPIC 10 REGULATION OF BODY FLUIDS

Biological Science, 4th ed.
Chap. 14, pp. 352–377
Elements of Biological Science, 3rd ed.
Chap. 13, pp. 220–235

Regulation of the osmotic concentration of body fluids is of little importance to the lower marine invertebrates, whose tissues and body fluids are isosmotic to sea water, and to a majority of plants, which can tolerate wide fluctuations in the composition and concentration of body fluids. For most other organisms, the ability to maintain a chemically stable and constant internal environment, usually quite different from the external environment, is absolutely essential to the integrity of the organism and its constituent cells.

As you have already learned, a cell can exercise a substantial degree of control over its internal environment. In part, this regulation is the passive result of physical characteristics of the cell membrane. To a remarkable extent, however, control is exercised through energy-requiring active transport, in which the cell may exclude certain substances and acquire others against normal concentration gradients or at a rate greater than diffusion would allow. Cells themselves require a reasonably constant environment, and we are, therefore, not surprised to find that natural selection has equipped organisms with a variety of devices to regulate the chemistry of body fluids. Many of these devices exploit the ability of cells to carry out active transport.

In this laboratory you will examine the excretory system of the rat, a representative regulatory system. Because the reproductive system of many vertebrates is closely associated with the excretory system, we shall use this opportunity to study the reproductive system of the rat as well. In the following laboratory, we shall look experimentally at active transport in a living mammal.

One of the principal requirements of animals is a means whereby the by-products of metabolism can be removed from the body. The principal wastes are carbon dioxide, from oxidative metabolism, and ammonia or some other form of nitrogenous waste, from protein deamination; both are toxic. You have already seen that carbon dioxide is discarded through diffusion from some type of gas-exchange surface. In the case of nitrogenous waste, excretion is a more serious problem since ammonia, the primary product, is highly toxic.

Some small animals can discard ammonia directly by diffusion from the body surface. Others must resort to excreting it in suitably dilute solution, by converting it into some less toxic form, such as urea or uric acid, or by combining these methods. In the great majority of animals, metabolic wastes are poured from individual cells directly into the body fluids; these fluids are then processed by special excretory organs, which release the wastes into the environment. Since most or all of the body fluids are processed by such devices, it is not surprising to find that these organs have often been endowed with the ability to retain selectively those materials useful to the organism, and thus to serve as fluid regulatory systems as well as excretory systems.

EXCRETORY AND REPRODUCTIVE SYSTEMS OF VERTEBRATES

Since the excretory and reproductive systems often develop together and are otherwise closely associated in vertebrates, the structures are often collectively termed the **urogenital system.**

Examining the rat's urogenital system

1. In your rat, locate the large paired **kidneys,** reddish organs attached firmly to the dorsal wall of the peritoneal cavity (Figure 10-1). Carefully remove the fat and peritoneum from the dorsal wall. This will allow you to see the kidneys, blood vessels, and reproductive organs more clearly. Locate the **adrenal gland,** a narrow, light-colored body, embedded in fat, lying along the anteriomedial portion of the kidney. This gland is part of the endocrine system, but its position makes it convenient to see at this time. What is the function of the adrenal gland?

2. At the mid-point of the medial side of the kidney is a depression where blood vessels are attached. Locate the renal artery and renal vein, tracing the former to its origin in the aorta and the latter to its junction with the posterior vena cava. You will note, also, that the **ureters** leave the kidneys and run posteriorly to the **urinary bladder.**

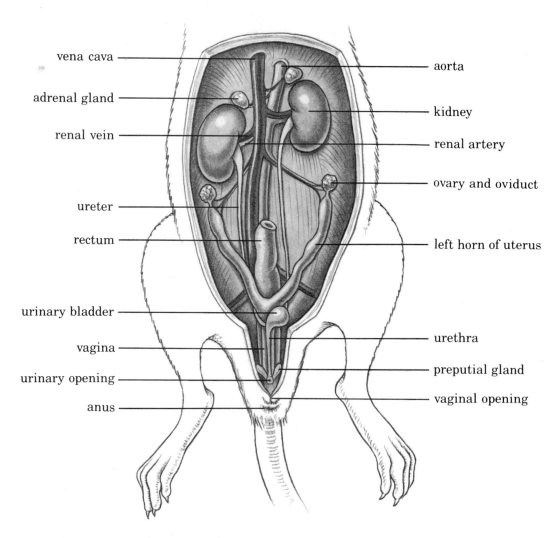

Fig. 10-1. Ventral dissection of the female rat urogenital system.

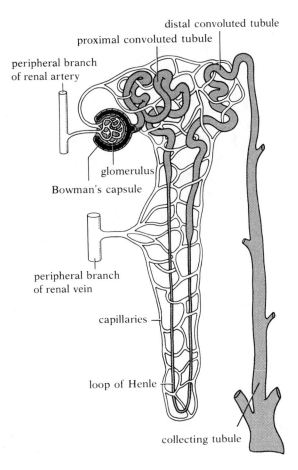

distal convoluted tubule

proximal convoluted tubule

peripheral branch
of renal artery

glomerulus

Bowman's capsule

peripheral branch
of renal vein

capillaries

loop of Henle

collecting tubule

Fig. 10-2. The human nephron. Each human kidney contains approximately 30 to 50 km of nephrons. The functioning of the nephron is explained in the text.

3. As you recall, the vertebrate kidney is very closely associated with a closed circulatory system. Each functional unit or **nephron** of the kidney is composed of a **capsule** enclosing a mass of capillary loops, the **glomerulus**, and a long tubule divided into three relatively distinct regions: the proximal convoluted tubule, the loop of Henle, and the distal convoluted tubule (Figure 10-2). A second capillary network is associated with the tubule. The distal tubules of different nephrons empty into common collecting ducts, which join, rejoin, and finally empty into the **renal pelvis**, a collecting area. The **cortex** of the kidney (the outer part, where the capsules are located), the **medulla** (the inner part through which tubules and collecting ducts pass), and the renal pelvis can all be seen clearly in a longitudinal section of the kidney. Remove one of the kidneys and section it longitudinally, or examine one on demonstration. Compare it to Figure 10-3.

Because the nephron is closely associated with the circulatory system, nephrons need not be distributed throughout the body (as is the case with the nephridia of earthworms, for example), but can be grouped together in the compact kidneys.

4. The **urethra**, which arises from the bladder medially between the ureters, runs posteriorly parallel to the rectum. It carries urine to the exterior. Follow the urethra until it passes from view into the ring formed by the pelvic girdle.

5. We shall now examine the reproductive system. As your dissection

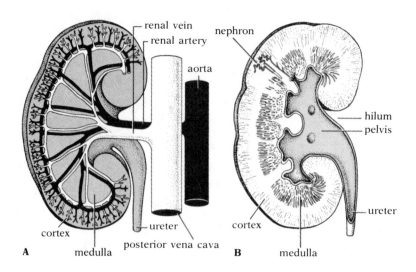

Fig. 10-3. Sections of the human kidney. Left: Kidney's blood supply; right: large renal pelvis into which numerous collecting tubules empty.

proceeds, compare the structures in the rat with the human reproductive organs illustrated in your text. Although we will not discuss the physiology of reproduction at this time, make certain that you understand the function of each structure studied, and review the hormonal control of reproduction in your text.

Female urogenital system

If your animal is female, locate the reproductive structures described in this section, with reference to Figure 10-1. If your specimen is male, examine another student's female specimen. Each student is responsible for knowing the reproductive structures of both sexes.

1. Locate the paired female gonads, the **ovaries**, small bodies suspended from the peritoneal wall in mesenteries posterior to the kidneys. Examine one ovary closely, and note the small, short, coiled **oviduct**, sometimes called the Fallopian tube. The oviduct is not attached directly to the ovary, but ends in an open, funnel-shaped structure that partially encloses the ovary. It is visible only with the dissecting microscope. When ovulation takes place, the eggs are carried by ciliary currents into the mouth of the funnel and pass down the oviduct to the uterus.

2. In your rat, most of the **uterus**, or womb, is divided into two sections called **uterine horns**. At its anterior end, each of the uterine horns is continuous with an oviduct; at their posterior ends the uterine horns join to form the smaller, median **body** of the uterus. Be careful not to confuse the uterine horns with the oviducts—the latter are much smaller and are located very close to the ovaries. (In the human female, the uterus is a single, median structure.)

3. The body of the uterus is continuous posteriorly with the **vagina**, slightly larger in diameter than the uterus and delimited from it by a slight constriction, the **cervix**. The vagina, lying between the rectum and the urethra, disappears from view into the ring of the pelvic girdle.

4. After you have identified the urethra, vagina, and rectum as they pass into the ring of the pelvic girdle, cut through the skin and muscle on each side of the urethra, and then cut very carefully through the bone of the pelvic girdle. With your fingers, spread the cut edges apart, and use blunt dissecting instruments to separate connective tissue from the vagina and urethra. Then follow these two ducts posteriorly to their separate external openings. The excretory and reproductive systems are completely independent of one another. Such independence of the two systems is relatively rare in vertebrates and is distinctive of female mammals.

The **clitoris** is associated with the urinary opening in the female rat, but in humans the two are separate. What is the function of the clitoris, and what is its homologue in a male mammal?

The preputial glands, which open into the vagina, are located on each side of the urethra. These glands are the female homologue of the male preputial glands.

Male urogenital system

The structures described in this section refer only to the male. If your specimen is female, examine another student's male specimen. Refer to Figure 10-4.

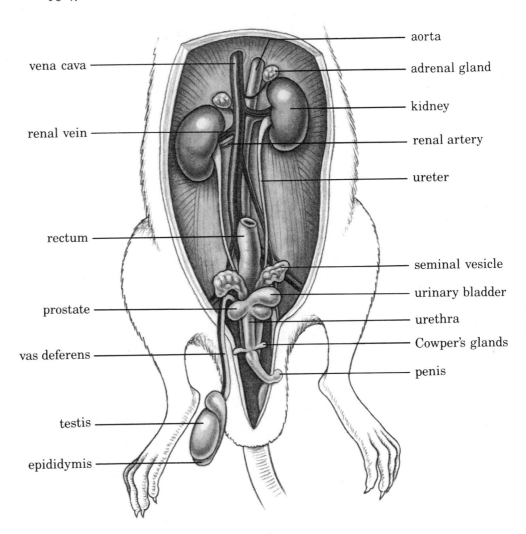

Fig. 10-4. Ventral dissection of the male rat urogenital system.

Locate the **scrotal sacs** on the exterior of the body, ventral to the anus. Each of the scrotal sacs contains a testis, surrounded by connective tissue, muscle, and skin. During fetal development, each testis first develops in the peritoneal cavity near the kidney, in a position similar to that occupied permanently by the ovaries in the female. As the fetus matures, however, the testes move posteriorly, or descend, each passing through the inguinal canal and into the scrotum. The inguinal canal remains through adulthood and permits the retraction of the testes into the abdominal cavity during nonbreeding periods. In some other mammals, including humans, the testes descend permanently into the scrotal sacs.

1. Cut open one of the scrotal sacs, and identify the testis. The testes produce sperm and testosterone, a hormone responsible for the production of secondary sex characteristics. Carefully clean away all the surrounding tissue (but do *not* cut any structures running anteriorly from the scrotal sac), and cut through the muscles ventral to the inguinal canal so that the full length of the scrotum, testis, and **spermatic cord**, a prominent cord extending from the anterior view of the scrotum well into the abdominal cavity, is exposed to view. The spermatic cord contains the blood vessels, a testicular nerve, and the vas deferens, or sperm duct.

2. Located on or around the testis is the **epididymis**, a very tightly coiled tubule, generally lighter in color than the testis itself. The coils of the epididymis temporarily store sperm cells, which are received from the testis through a large number of tiny collecting tubules, too small to see in your dissection. Sperm enter from the epididymis into the **vas deferens**. Trace the vas deferens anteriorly as it loops over the ureter and turns posteriorly to join with the other vas deferens and then with the urethra.

From studying comparative anatomy and embryology of amphibians and mammals, we know that the mammalian ureter is a new duct and that the archinephric or Wolffian duct of amphibians, through evolution, has become the epididymis and vas deferens.

3. Remove the muscles from the pelvic girdle posterior to the urinary bladder. Carefully cut through the pelvic girdle, and spread open the pelvic region. At the point where the united vas deferentia (the plural form of "vas deferens") enter the urethra, locate a pair of large, lobulated glands: the **seminal vesicles**. The **prostate glands** are located on each side of the urethra, just at the base of the urinary bladder.

4. The urethra extends from the base of the urinary bladder into the penis. At the base of the penis, identify another pair of accessory sex glands: **Cowper's glands**, or bulbo-urethral glands. These glands, along with the seminal vesicles and prostate, secrete fluids that make up part of the semen. The urethra functions as a urogenital canal—i.e., a common duct for urine and semen. During the passage of seminal fluid, the duct does not carry urine. It is cleansed by secretions that precede the semen, and any acid that remains is neutralized by substances in the semen.

In many lower vertebrates the excretory and reproductive systems are even more intimately united. The ureters do not empty into the urinary bladder as in mammals, but join the dorsal surface of the **cloaca**, a posterior chamber of the digestive tract. Urine thus flows from the ureter into the cloaca and out the opening leading from the cloaca to the exterior. What other types of vertebrates show a cloaca rather than separate urinary, genital, and digestive ducts?

QUESTIONS FOR FURTHER THOUGHT

1. What is the function of testis descent in a mammal? Why should the testis not descend in a frog?

2. What happens to ammonia and other nitrogenous wastes after they are excreted by animals?

3. In male mammals and amphibians, the urinary and genital systems share common ducts along much of their length. What selective advantages would there be to having separate tracts in the females?

4. What structure in the rat is functionally comparable to the Wolffian duct in the frog?

5. What do kidneys and nephridia have in common?

6. Contrast osmoregulatory adaptations of fresh-water bony fish with those of marine bony fish and with those of sharks.

7. Discuss the process of urine formation in man, explaining:

    a. The source of urea and where the urea content of the blood is highest.

    b. The roles of filtration and reabsorption, and where each of these processes occurs.

Biological Science, 4th ed.
Chap. 4, pp. 103–105; Chap. 14, pp. 375–377
Elements of Biological Science, 3rd ed.
Chap. 4, pp. 75–77

For those situations when diffusion would be too slow or when substances must move against a concentration gradient, natural selection has exploited the ability of cells to carry out active transport. As you already know, energy-requiring transport processes are extremely widespread in organisms, ranging from mineral uptake in roots and selective reabsorption in the kidney tubules, to the "sodium pump" that maintains the electrochemical potential difference necessary for the proper functioning of nerves and muscle fibers. The mechanism of active transport of metabolites into the cell always involves the expenditure of ATP. Research has shown that sodium-ion (Na^+)-linked carrier proteins are also necessary, and that these carrier proteins can discriminate between different molecules. Amino acids, for example, exist in two configurations, differing only slightly in the spatial arrangement of atoms; the mouse intestine, in which you will presently study the transport of glucose, actively transports into the blood one of these forms (the L-configuration) but not the other (D-configuration).

The absorptive surface of the small intestine of the mouse, like that of many other animals, is greatly augmented by large numbers of **villi**, fingerlike projections of the intestinal wall. Each villus, furthermore, is lined with cells that constitute the intestinal **mucosa**, and each mucosal cell (Figure 11-1) further increases the surface area for absorption with numerous tiny projec-

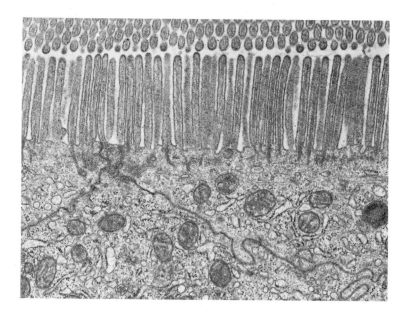

Fig. 11-1. Electron micrograph of mouse intestine showing microvilli projecting from the border of several epithelial cells; notice the numerous mitochondria. *(Courtesy E. W. Strauss, Brown University.)*

tions (**microvilli**) from its surface. Almost all the absorption of food materials, including glucose, amino acids, fatty acids, and glycerol, occurs across the membranes of these mucosal cells. Behind the mucosa, there are numerous capillaries and lymph vessels. Food is carried in the blood to the liver, where it may be stored and released to the body as required.

If the transport of digestive products depended only upon diffusion, their level in the blood would never exceed their level in the small intestine and much food would be lost. In fact, the concentration of glucose and other digestive products in the blood may substantially exceed the intestinal concentration. The difference is brought about and maintained by active transport.

The procedure you will use today was designed by T. H. Wilson of the Harvard Medical School and Dr. G. Wiseman of the University of Sheffield (England). It involves cutting the mouse intestine into sections of workable length, everting the sections so that the mucosal (external) side is outermost, and then filling the sections with a solution of glucose in Krebs-Ringer solution. The sacs are then incubated in the same solution. After incubation, the levels of glucose inside and outside the sacs are measured and compared. During incubation, active transport will have moved glucose across the mucosa into the new "lumen" of the intestinal sacs. This basic procedure may be modified in various ways to demonstrate a requirement for ATP or for oxygen, or to demonstrate the selective transport of the L-amino acids.

Read through the entire experiment carefully before you begin to work. In Figure 11-2, identify all the equipment as you read through the directions. During the incubation, make sure that the equipment for making glucose analyses is operating properly. If the instructor does not demonstrate the method of analysis, carry out several trial runs on samples of the glucose Krebs-Ringer solution and on dilutions to make sure that you understand how the analyses work. Students should work in pairs throughout the experiment; in the first two sections, one student can operate while the other makes sure that instruments, solutions, and other equipment are made ready for the operator.

TISSUE PREPARATION FROM THE EXPERIMENTAL ANIMAL

Each team will be provided with a freshly killed animal. Open the abdomen with a mid-line incision, and extend it laterally to give yourself more working room. Quickly identify the small intestine and its upper and lower limits (the pylorus and the region near the caecum).

Cut part way across the duodenum near the stomach; insert the cannula and tie it firmly in place with thread. Run warm 0.9% saline solution from the burette through the cannula into the small intestine. Watch the flow through the intestine, and, when fluid appears in the caecum, stop the flow of saline. Then cut across the intestine just above the caecum, and allow more saline to rinse through the gut. Lifting the intestine gently near the caecal end, strip the gut of its mesentery up to the end of the duodenum. Remove the intestine, noting its polarity, and place it in a petri dish of *warm* saline. Since different sections of the intestine demonstrate different transport capabilities, it is important to keep track of the polarity of the intestine

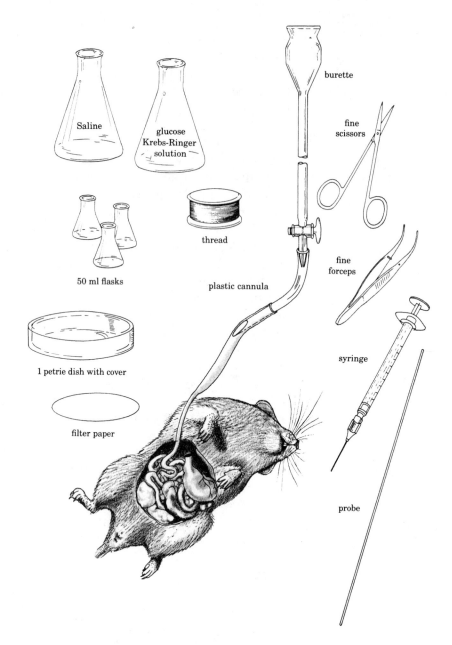

Fig. 11-2. Mouse dissection and equipment for the active-transport study. The intestine has been cut across the duodenum, and a plastic cannula has been inserted.

throughout the experiment. Wrap the animal in a paper towel, and set it aside for a quick review of the internal anatomy when you have time later in the period.

PREPARATION OF INTESTINAL SACS

Your instructor will show you a film on how the intestinal sacs are prepared, or he or she may demonstrate the technique. After you have seen the demonstration, you will be ready to follow the outline below.

Preparing intestinal sacs   1. Blot the *lowest* centimeter of the intestine (that farthest from the stomach) on filter paper, and push it back through the lumen with your special probe. When it appears at the other end, advance the tissue along the probe until the entire gut is everted with the serosal (internal) surface exposed. Remove the tissue by pulling it carefully off the probe, and place it at once in fresh, warm saline.

2. The gut is now to be cut into three or four sections, depending on the length of intestine available, and each section is to be made into a sac. Tightly tie a thread around the intestine about 3 to 5 cm from the duodenal end (about one-quarter the length of the whole gut), and cut the tissue distal to the thread. Repeat this procedure until you have several such sections, each closed at one end by a thread ligature.

3. Each section will now be filled with the glucose Krebs-Ringer solution and tied off. It is important that the sacs be *slightly distended* (not limp) with solution, so follow the directions carefully.

Holding a short section of tissue by its thread, blot it gently on filter paper to remove adhering fluid, and place the tissue on an inverted petri dish (Figure 11-3). Fill a 1-ml syringe with glucose Krebs-Ringer solution, and

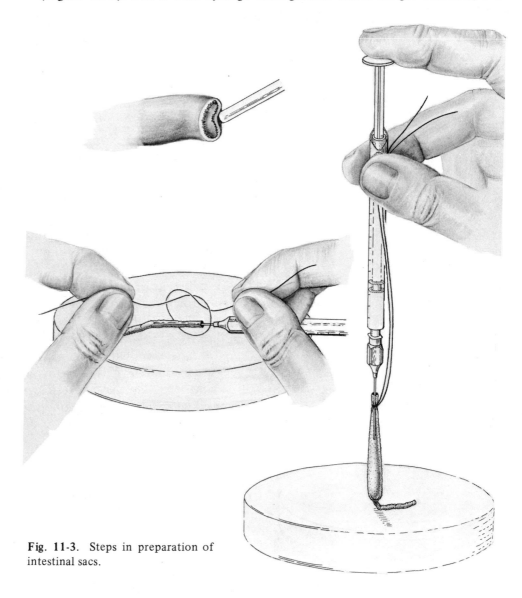

Fig. 11-3. Steps in preparation of intestinal sacs.

loosely tie a thread over the barrel. Holding the open end of the tissue with forceps, introduce the blunt needle into the sac. Rest the syringe on the edge of the petri dish, pull the thread over the tissue and needle, and draw the thread tight.

With one hand, hold both the syringe barrel and the two ends of the ligature, and lift the sac off the dish, allowing the tissue to hang freely. Inject the glucose Krebs-Ringer solution until the sac is filled and slightly distended. Do not allow any air to remain in the sac. Slide the sac off the needle, and pull the thread tight. Trim any large "ends" of tissue or thread, and place the sac in a 50-ml Erlenmeyer flask containing 5 ml of glucose Krebs-Ringer solution.

Repeat this procedure until the other sacs have been prepared. Each should be placed in a separate flask.

You have now prepared several sacs, each containing an isosmotic solution of salts and 10mM glucose. Each has been placed in a flask of the same solution. The osmotic concentration difference between the mucosal (outer) side of the sac and serosal (inner) side of the sac is thus zero. If active transport occurs, this relationship will be altered and can be detected by analyzing the glucose concentration on each side of the sac.

4. To test the requirement in active transport for ATP, add a drop of dinitrophenol, an inhibitor of cellular respiration, to one of the flasks, and make sure it is labeled so that it can be clearly differentiated from the others.

5. Each of the flasks should now be aerated and incubated according to one of the methods below.

AERATION AND INCUBATION

Option A   Fit each flask with a two-hole stopper. Connect one of the openings to an air line, and bubble air through the flask at room temperature for one and a half hours. The air stream should provide some circulation of the flask fluid but should be gentle enough so that the tissue will not be damaged.

Option B   Fit each flask with a two-hole stopper, and run oxygen from a small tank into one of the two holes for thirty seconds under pressure to produce a flow rate of 2 liters per minute. Promptly stopper the flask or plug the holes, and incubate the flask at 37°C with shaking for one hour or at room temperature with shaking for one and a half hours.

While the sacs are incubating, gather the materials for the glucose analyses, carry out trial runs, examine the mouse to review the internal anatomy of a mammal, and follow any other directions given by the instructor at the beginning of the period.

COLLECTION OF SAMPLES

After incubation, remove the stoppers from each flask, and pour the solution from each one into a separate test tube, which you have previously labeled.

With forceps, pick up each sac by one of its threads. Touch the dangling end *once* with filter paper to remove excess fluid, and hold the end of the sac just inside the lip of a small test tube. Carefully cut the end with scissors, and drain the sac contents for fifteen to thirty seconds. Keep the samples from each flask separate, and be especially careful to label those tubes used with the dinitrophenol-treated tissue.

GLUCOSE ANALYSIS

Compare the glucose concentrations of sac contents and incubation solution by one of the following methods, as directed by your instructor.

Option A   If a colorimeter is available, the glucose-oxidase method of analysis may be used. In the presence of this enzyme, glucose is oxidized to gluconic acid and hydrogen peroxide. The peroxide reacts with a hydrogen donor in the presence of peroxidase to produce a yellow-orange product, the concentration of which may be estimated colorimetrically.

1. Pipette 0.10 ml of final serosal (external) solution, final mucosal (internal) solution, and initial glucose Krebs-Ringer solution into separate tubes.

2. Add 0.9 ml of water to each tube. Use 1.0 ml of water in a fourth tube as a "water blank." The initial solution, containing 10mM glucose, may be used as a standard.

3. Now, add 5.0 ml of the glucose oxidase reagent to each successive tube, making the additions one minute apart.

4. Exactly ten minutes after the reagent was added to the first tube, add 1 drop of 4 N hydrochloric acid to each, at one-minute intervals, until all tubes have been completed.

5. After five minutes, the color may be read in a colorimeter at a wavelength of 400 nm. The colored product is stable for a few hours.

Option B   If a colorimeter is not available, less accurate estimations of concentrations may be made by reacting a small amount of each solution to be tested with Benedict's reagent. The results are compared with known glucose standards.

In the Benedict's reaction, blue cupric ion (Cu^{++}) is reduced to red cuprous ion (Cu^{+}) by the aldehyde group of the sugar. The cuprous ions react with hydroxyl ions to yield cuprous oxide, a red product. The color obtained varies with the amount of glucose originally present: a clear blue solution indicates no glucose; a deep red indicates high glucose concentration.

To test a solution for glucose concentration, place 3 drops of the solution in a small test tube, add 2 ml of Benedict's reagent, and place the tube, properly labeled, in a boiling water bath for five minutes. Then observe the color.

Follow this procedure for a sample of the solution inside each intestinal sac, a sample of the external solution for each, a sample of the initial external solution, and "standard" comparison tubes containing 0, 1, 2, and 3 drops of the standard glucose Krebs-Ringer solution.

REPORTING THE RESULTS

Organize the results into a table, and prepare diagrams, appropriately labeled to indicate the direction and extent of glucose transport in each sac.

ADDITIONAL EXPERIMENTS

If time and materials are available, this experiment can be more quantitative or can be modified to demonstrate the hydrolysis of starch in the upper jejunum (versus the ileum) and to demonstrate selective transport of L-tyrosine. Your instructor will provide you with the details.

TOPIC 12 VERTEBRATE COORDINATION

Biological Science, 4th ed.
Chap. 18, pp. 451–490
Elements of Biological Science, 3rd ed.
Chap. 17, pp. 284–305

From your previous work with the rat, it should be clear that the vertebrate body is an association of extremely complex and intricate parts. Some of these parts, such as the skeleton and its associated muscles, expedite voluntary movements; others, such as the respiratory and circulatory systems, carry on their functions involuntarily. While each system of the body has a specific contribution to make, all must function together as a coordinated whole. Such integration depends upon the rapid exchange of information between widely separated regions of the body, and in this process the nervous system plays a vital role. In this unit you will examine the development and structure of the vertebrate nervous system and conduct some experiments to determine the role of the central nervous system in communication within the organism.

STRUCTURE OF THE NERVOUS SYSTEM

For purposes of discussion, the nervous system is conveniently divided into three major sections. The **central nervous system**, or CNS, is composed of the brain and spinal cord; the **peripheral nervous system** constitutes all nervous connections outside the CNS; and the **autonomic nervous system**, a subdivision of the peripheral, regulates involuntary functions.

Central Nervous System   Examine the skull and spinal column of the vertebrate skeleton on demonstration. The brain lies inside the cavity of the skull, and the spinal cord runs from the base of the brain down inside the spinal column. Paired **spinal nerves**, containing nerve fibers that bring information from and carry it to all parts of the body, leave the spinal cord at intervals through openings in the **vertebrae**.

Development of the brain

1. Examine on demonstration a two-day-old chick embryo, referring to Figure 12-1. The brain is its most prominent structure and has three distinct parts, separated by constrictions. The **midbrain** is the most anterior because the brain has doubled back on itself during development. The **forebrain** can be identified posterior to the midbrain and adjacent to the developing eye, the optic cup. The **hindbrain,** consisting of several lobes, blends into the spinal cord, which can be traced posteriorly in the longitudinal axis of the body. These parts of the brain differ in anatomy and in function in different vertebrate groups, but early development is similar in all vertebrates. Which of the parts becomes dominant in human beings?

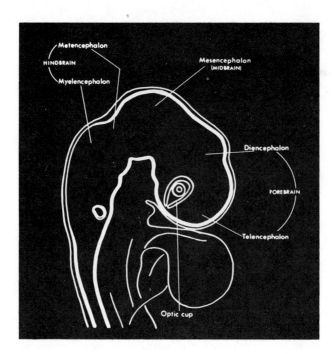

Fig. 12-1. Outline of forty-eight-hour-old chick brain. *(Courtesy Carolina Biological Supply Company.)*

Mammalian brain

2. You will be provided with a mammalian brain in a container of water. Although it was previously fixed in 10% formalin, it has been thoroughly rinsed by your instructor. Be very careful in handling your specimen since it is extremely delicate.

As you examine this mammalian brain, compare it with Figure 12-2, illustrating adult brains of other vertebrates, and with Figure 12-3, showing a longitudinal section through the mammalian brain.

Apart from tracts that conduct impulses through its various regions, the brain contains centers, each of which dominates regulation of one or more major activities in the body.

    a. The most posterior part of the hindbrain, part of which may be missing in your specimen, is the **medulla**, which controls a number of important reflexes, such as coughing, sneezing, blinking, and swallowing, and contains centers for the parasympathetic control of respiration, heartbeat, visceral movement, and glandular secretion. Notice numerous nerves leaving the medulla. The **cerebellum** is located just anterior to the medulla on the dorsal side of the hindbrain. It maintains equilibrium of the body by correlating information from kinesthetic ("movement detector"), touch, pressure, and static receptors.

    b. The small midbrain, front part of the brain "stem," is best seen in the longitudinal section, Figure 12-3. It processes information from the eye and ear.

    c. The **pituitary gland**, which exercises control over most of the other glands in the body, hangs from a portion of the forebrain called the **hypothalamus** (the pituitary may have been damaged or removed when the brain was prepared for preservation). The largest and most anterior part of the forebrain is the **cerebrum**, consisting of halves that have almost completely overgrown the rest of the brain. Convolutions of the cerebrum provide a great surface area for neuron cell bodies, located in the outermost layer or **cortex** of the cerebrum. What is the function of the cerebrum in a mammal? Is it the same in other vertebrates? Beneath the front edge of the cerebrum, locate the **olfactory bulb**. What is its function?

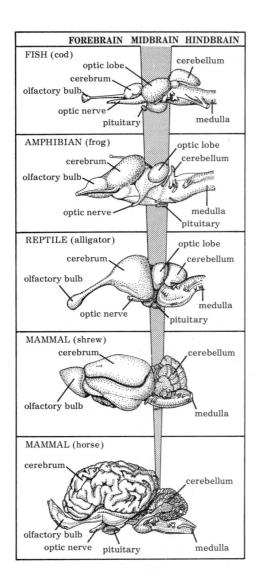

Fig. 12-2. Evolutionary change in the relative size of the midbrain and forebrain in vertebrates. In the sequence from fish through amphibian, reptile, and primitive mammal (shrew) to advanced mammal (horse), the relative size of the midbrain markedly decreases, while the forebrain expands enormously. *(Modified in part from A. S. Romer,* The Vertebrate Body, *Saunders, 1962, and in part from G. G. Simpson, C. S. Pittendrigh, and L. H. Tiffany,* Life: An Introduction to Biology, *copyright © 1957 by Harcourt, Brace & World, Inc. Used by permission of the publishers.)*

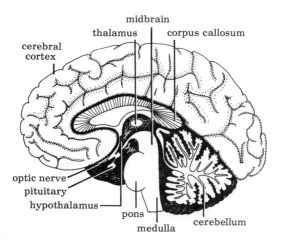

Fig. 12-3. The human brain in sagittal section. *(Modified in part from S. W. Ransom and S. L. Clark,* The Anatomy of the Nervous System, *Saunders, 1959.)*

Spinal cord

3. a. Examine a dissected spinal cord on demonstration. Note the shape of the cord, and identify two slightly enlarged regions—the cervical and lumbar enlargements—reflecting the increase in neurons required to supply the fore- and hindlimb muscles.

b. Dissect the skin, muscle, and bone away from a portion of your rat's spinal cord (by using a razor blade to shave away tissue down to the cord, and then by blunt dissection). Do your dissection under a dissecting microscope, and identify dorsal and ventral roots of a spinal nerve (see the description in "Response to Noxious Stimuli," below).

c. Also examine a cross section of the spinal cord. Locate the points of attachment of spinal nerves, and study one of them carefully, noting dorsal and ventral roots of the nerve. Which root carries afferent impulses, and which efferent impulses? The inner part of the cord is an H-shaped area containing gray matter, the nerve cell bodies. Outside the gray matter is the white matter, composed of nerve fibers extending lengthwise and also branching off into spinal nerves. In the center of the cord is a small **central canal**, which contains spinal fluid and is continuous anteriorly with the **ventricles**, fluid-filled cavities in the brain.

The dorsal part of the gray matter in the spinal cord contains cell bodies of association neurons, and the ventral part contains the cell bodies of motor neurons.

Neurons in the CNS

4. Examine your slide of giant multipolar neurons. These slides are made by staining smears of cells from the ventral horn of the spinal cord of an ox. Locate one of the large bodies, and note the many processes running from it. Make a sketch of one of these neurons.

Peripheral Nervous System

Spinal nerves

1. Twenty to thirty pairs (or more, depending upon the vertebrate in question) of spinal nerves leave the spinal cord between the vertebrae. Each of them has a **dorsal root** carrying sensory (afferent) impulses and a **ventral root** carrying motor (efferent) impulses. These roots can be located on a slide of a cross section or may be seen in dissection, as noted previously. The bulge on the dorsal root—the dorsal-root ganglion—contains the cell bodies of sensory neurons. Wherever nerve cell bodies aggregate outside the CNS in this manner, we speak of them as forming a **ganglion**.

Other nerves

2. The specialized, impulse-conducting cells that form the nervous system are called **neurons** or nerve fibers. Figure 12-4 illustrates three major types of neurons and will remind you that each consists of a **cell body** and variously branched processes called **axons** and **dendrites**. Dendrites are processes that conduct impulses toward the cell body, while axons usually conduct impulses away from the cell body. Neurons commonly have a single axon and many dendrites, but there are many exceptions to that rule.

Sensory neurons carry impulses from the organs of special sense to the CNS; motor neurons conduct impulses toward the muscles and other effector organs; and association neurons conduct impulses between the sensory and motor fibers, usually within the CNS.

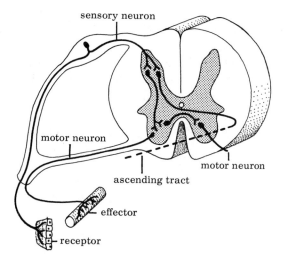

Fig. 12-4. Diagram of a reflex arc with association neurons. The sensory-neuron synapses with several association neurons in the gray matter of the spinal cord. Some of these association neurons may synapse directly with motor neurons on the same side, but some cross to the other side of the cord and there synapse with the other motor neurons and with additional association neurons that run in ascending tracts through the cord to the brain.

Nerves are collections of neurons into tracts of substantial size.
   a. Examine a slide showing a cross section through a peripheral nerve. Note the many separate fibers (neurons) of which it is composed, and the dark myelin sheath around each fiber. What is the function of the myelin sheath?
   b. Examine a slide showing the **neuromuscular junction**, that area where a neuron and the group of muscle fibers innervated by it connect. The axon of a motor neuron divides into a number of fine terminal branches, each ending as a **motor end plate** on the muscle fiber. Transmission of the impulse across the cleft of the neuromuscular junction is mediated by chemicals such as acetylcholine and adrenalin.

THE REFLEX ARC

Some stimuli cause rapid responses that come about involuntarily and often without the intervention of the higher centers of the brain, although in many cases those centers are informed and may modify the response. These automatic responses are called **reflexes**. The sensory impulses that initiate a spinal reflex enter the spinal cord through the dorsal root. Within the cord, they usually synapse with association neurons, which, in turn, transmit the impulses to motor neurons and then to some effector, such as a muscle. The whole pathway, from sense organ through the CNS to effector, is called a **reflex arc** (Figure 12-4).

As you examine some reflexes in a frog, do not forget that spinal reflexes, though they do not *require* the intervention of the brain, may nevertheless be modified by it and that most reflexes are not nearly as simple as we have outlined above.

(*Note:* The following exercises require several frogs per student group. If frogs are in short supply, students can reverse the order of the experiments and can work in groups of up to four.)

Pith a frog in the usual manner, but only anteriorly, leaving the spinal cord intact. (Directions for pithing are in Appendix 3.) This procedure destroys the brain and separates it from the spinal cord. Presumably, the frog has no sensations and can initiate no voluntary actions.

Posture

Coordination Reflexes   Place the frog on a table. Does it assume, or try to assume, a squatting posture? Do the hind legs position themselves under the body when pulled straight?

Jumping

Try to stimulate the frog to jump. Do you have any success?

Righting

Place the frog on its back. Does it make any attempt to right itself?

Balance

Place the frog on a board. Tilt the board. Does the frog respond at all?

Swimming

Does the frog swim when placed in deep water? Does it make any attempt to get out of the water?

Response to noxious stimuli

Hang the frog from a ring stand (Figure 12-5) so that the legs are allowed to hang freely. One way of accomplishing this is to put a safety pin through the frog's lower jaw, using a string to attach the pin to the ring-stand clamp.

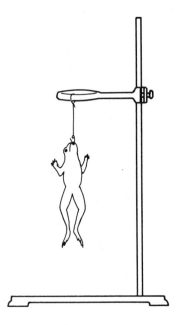

Fig. 12-5.   Manner in which a frog is hung from a ring stand.

1. Use forceps to pinch one of the toes. What is the result? Try increasing the severity of the stimulus. What are the effects?

2. Pinch various parts of the body. Are the responses localized with reference to the stimulus? Are these responses predictable (i.e., will the same stimulus always yield the same response)?

3. Apply some 10% acetic acid to one of the toes with a cotton swab. Are the responses the same as the ones from pinching? Try 25% acid, being sure to rinse the specimen well between renewed applications by dipping the toe in water. What are the results?

4. Soak a small piece of filter paper (½ × ½ cm) with 25% acetic acid, and apply it middorsally to the skin of the pelvic region. What are the responses? Try the same technique on other parts of the body, washing off the acid between applications by using a cotton swab soaked in water. Can these movements be accurately predicted? What happens when one of the responses is interfered with (as, for example, by holding one of the legs)?

Control by the Cerebellum and Medulla Place a refrigerated frog on crushed ice until the frog no longer moves about. Locate the tympanic membranes (the brown, circular membranes behind the eyes). What is their function? Using the anterior margin of the two membranes as a guideline, cut off the anterior portion of the skull with a pair of heavy scissors. Leave the lower jaw attached. You have removed the fore- and midbrain as shown in Figure 12-6.

After ten or fifteen minutes, introduce the frog to the reflex situations discussed in "Coordination Reflexes" and "Response to Noxious Stimuli," above, and record the results. What can you conclude about the site of control of coordination reflexes? Of the site of control of the reflexes discussed in "Response to Noxious Stimuli," above?

Fig. 12-6. Frog, showing cut to remove the fore- and midbrain.

Functions of the Midbrain and Cerebrum Place an unoperated frog in the same situations as the other two frogs (except those dealing with noxious stimuli), and determine the general additional functions of the midbrain and cerebral hemispheres. Are the responses always predictable?

SENSE ORGANS

Neurons can be stimulated to "fire" by almost any kind of treatment: electric shock, chemical agents, cutting, pinching, etc. Under normal conditions, sensory neurons are activated by the sense organ to which they are connected. A sense organ is itself activated by some type of environmental change in energy, a **stimulus,** and it somehow causes attached neurons to fire when it is so stimulated. The neurons then transmit the information to the CNS. The sense organ itself does not interpret stimuli as sensations, although it may subject the stimuli to filtering or other selective processes before information is passed on to the CNS. Interpretation takes place, and response is initiated in the CNS. Interpretation is a function of the region of the CNS to which the impulses are transmitted, not of the type of stimulus or the type of receptor.

We shall investigate one important property of sense organs and shall dissect and examine a representative receptor.

Adaptation   Under conditions of continuous stimulation, most types of receptors cease to become excited, or their rate of transmission of impulses to associated neurons declines. This phenomenon, which helps to reduce the transmission of useless information to the CNS, is called **adaptation**. Proprioceptive sense organs—those concerned with the transmission of information about muscle tension, posture, etc.—are generally slow-adapting. If this were not so, we would not be able to stand up or to balance ourselves without continuous conscious effort. Exteroceptive sense organs—those that supply information about events at the body surface, such as heat, cold, touch, etc. —generally adapt very rapidly. If they did not, impulses from all parts of the body, mostly useless information, would impinge continuously upon the CNS.

Testing adaptation   Test the adaptation of heat and cold receptors in the following manner:

1. Set up three beakers, one with cold water, one with tolerably hot water, and one with water at room temperature.
2. Place one index finger in the cold water and the other index finger in the hot water. Report your initial sensations and sensations after several minutes. Before removing your fingers from the hot and cold water, read "Identifying the Stimulus Required for 'Hot' and 'Cold' Sensations," below.

Identifying the Stimulus Required for "Hot" and "Cold" Sensations   Move the two fingers together into the beaker of water at room temperature. Note that although the stimulus is superficially the same, the sensation in each finger is different. Are the stimuli really identical for each finger?

Receptors in the cold finger respond to heat gain, whereas receptors in the hot finger respond to heat loss. Does this experiment tell you whether the receptors for these stimuli are the same or different?

The Eye: A Representative Sense Organ

1. Obtain a sheep or cow eye, and place it in a dissecting pan. Locate the brownish stubs of muscles that controlled the position of the eyeball and the white **optic nerve**. With scissors, cut the eyeball in approximate halves, in a plane parallel to the front surface.
2. Excluding fat on the surface, note that most of the wall of the eye is composed of three layers (Figure 12-7):
   a. A white, tough outer **sclera**, specialized anteriorly as the transparent **cornea**. It may be milky because of preservation.
   b. The black **choroid**. It contains numerous blood vessels supplying the retina. (What is the function of the black color of the choroid?)
   c. The innermost, membranous gray **retina**, containing the light-sensitive cells.
3. Locate the biconvex **lens** lying in the interior of the eye toward the

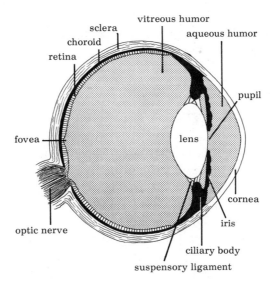

Fig. 12-7. Diagrammatic section through the human eye.

front. In life, the lens is transparent and acts as the focusing device. The cavity anterior to the lens is filled with **aqueous humor**, while the cavity between the lens and the retina is filled with the more viscous **vitreous humor**. What is the function of these fluids? Remove the lens and fluids from the eyeball. If you break the lens apart with your fingernail, you will find that it is composed of many fiberlike cells packed closely together.

4. Examine the choroid layer more closely. You will see that toward the front of the eyeball, near the lens, the choroid forms a thickened, ring-shaped, radially striated muscular area, the **ciliary body**, that curves inward to enclose the lens. The ciliary body is the point of attachment of the suspensory ligament, a thin membrane attached to the lens around its equator. During the process of accommodation, contraction of the muscle fibers in the ciliary body reduces the pull on the suspensory ligament and thus on the lens, allowing the naturally elastic lens to assume a more rounded shape, which brings nearby objects into better focus. Conversely, when the muscle fibers in the ciliary body relax, they elongate, thus exerting great pull on the suspensory ligament and on the lens, and pulling the lens into flatter shape, better adapted to distance vision. In older individuals, the lens often loses its original elasticity, and accommodation diminishes. Older persons then need corrective lenses for near vision.

5. Continue to follow the choroid layer, past the ciliary body and forward. It continues as the **iris,** which turns inward, leaving the wall of the eye and covering part of the front of the lens. The iris is perforated immediately in front of the lens; the hole in the iris is called the **pupil.** The iris has muscle fibers, some circular and others radially arranged. The iris can thus act as a diaphragm regulating the size of the pupil and, in turn, the amount of light admitted to the retina.

6. Examine the rear half of the eye more carefully. Note that the pale, loose retina is attached to the wall of the eye at the **blind spot,** where the sensory fibers from the retina join the optic nerve and where blood vessels pass into the eye. Beneath the rear portion of the retina, the choroid layer is covered by a shiny greenish-blue material. This is the **tapetum,** a reflecting area that sends incident light back through the retina, thus making more efficient use of the light. There is no tapetum in the human eye, but the

device is common in nocturnal terrestrial vertebrates and in fish that live far below the surface. It is the tapetum that causes the so-called "glowing" eyes of some animals at night.

ELECTRICAL CHANGES IN CONDUCTIVE TISSUE

An oscilloscope is an instrument capable of picking up electrical stimuli from an object and graphing them on a screen that closely resembles a television screen. Changes in the electrical input to the instrument are viewed as deflections of a light trace on the graphlike screen. An oscilloscope can be a valuable tool in studying the electrical changes that take place in a muscle or a nerve or another part of an organism under specified experimental conditions.

The instructor may carry out a demonstration of electrical changes occurring in nerve, muscle, or other tissues, and he or she will explain the peculiarities of the oscilloscope used.

QUESTIONS FOR FURTHER THOUGHT

1. What are some of the evolutionary tendencies in the organization of the nervous system from protozoans to mammals?
2. What is the function of the hypothalamus?
3. Explain polarization reversal.
4. Define and explain the role of the sodium-potassium pump.
5. Explain how an impulse is transmitted along an entire reflex arc, mentioning all the components of the arc and all the processes occurring in transmission.
6. Mention the different sense receptors in humans, and indicate the type of stimulus to which each receptor responds.
7. Using diagrams where necessary, describe the structure of the eye and its method of accommodation.

TOPIC 13 EFFECTORS

Biological Science, 4th ed.
Chap. 20, pp. 526–552
Elements of Biological Science, 3rd ed.
Chap. 19, pp. 329–345

One of the chief fascinations most of us find in animals is their endlessly varied capacity for movement. Animals in action in fight or flight are the most obvious and dramatic examples of movement, but almost all of an animal's activities—from chasing, chewing, and digesting food to communication with prospective sexual partners—are impelled by movements of one sort or another. In most animals, movements are brought about by muscles acting against some form of skeleton.

Muscles are **effectors,** biological devices that do things in response to stimuli, usually from the nervous system. There are many other kinds of effectors, such as nematocysts (in which phylum?), cilia, and glands, but muscles are the most common and intensively studied form. Some effectors may be "independent" (not subject to any control other than that built into themselves) or may be under the control of something other than the nervous system; but skeletal or voluntary muscles, which you will examine today, are almost exclusively under nervous control.

In this unit, you will investigate several different kinds of skeletons, their associated muscular effectors, and the physiology and biochemistry of muscular contraction.

SKELETAL SYSTEMS

One can immediately appreciate the importance of structural support in terrestrial organisms, but it is less obvious that a skeletal system is essential for rapid and efficient movement in any habitat. Arthropods and vertebrates— the two phyla of animals that have evolved the most effective locomotion— are also the two that have well-developed jointed skeletal systems. We shall examine representatives of these phyla and shall look briefly at one other type of skeleton: the hydrostatic skeleton.

Hydrostatic Skeleton: The Earthworm   Although one may not ordinarily think of the earthworm as having a skeleton, the action of body-wall muscles on the incompressible fluids within brings about very effective movement. We, therefore, describe this arrangement as a hydrostatic skeleton.

Examining the exoskeleton

1. Examine a cross section of the earthworm (*Lumbricus*), and locate in the body wall two sets of muscles (Figure 13-1): an inner, longitudinally oriented set, arranged in featherlike bundles, and an outer, circularly oriented set. The longitudinal muscle fibers are oriented parallel to the long axis of the body, and you will, therefore, see them in cross section; the circular

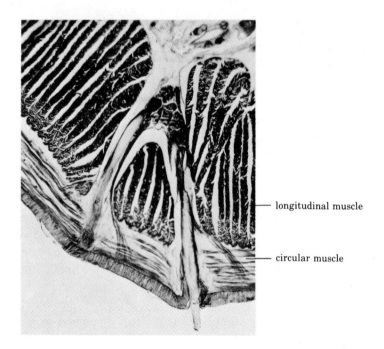

longitudinal muscle

circular muscle

Fig. 13-1. Cross section of an earthworm outer body wall. This photomicrograph shows longitudinal and circular muscles. *(Courtesy Ward's Natural Science Establishment, Inc., Rochester, New York.)*

fibers will be seen in longitudinal section, since they run in the same plane as the cross section of the body. What effect would contraction of the longitudinal muscles have on the shape of the worm? What effect would contraction of the circular muscles have on the shape?

Take this opportunity to review other structures in the cross section: intestine with typhlosole (what is its function?), dorsal blood vessel, ventral nerve cord, setae, and perhaps parts of nephridia.

2. Since the body is divided by membranes (septa) into compartments, and since each compartment (segment) contains relatively incompressible fluid, a compartment that contracts in diameter (by the action of which muscle set?) must simultaneously increase in length. Following this basic principle, the earthworm's fully segmented body is capable of a great variety of movements. Which movements occur depends upon the activity of the longitudinal and circular muscles in different parts of the body.

Obtain a pair of living earthworms. Place one on a wet paper towel in a pan and the other on a wet plate of glass. Observe carefully the movements of the body and the effectiveness of locomotion on these two substrates.

Forward movement is accomplished by a wave of contraction of the circular muscles accompanied by relaxation of the longitudinal muscles and, therefore, elongation of the body. Which way along the body does the contraction wave pass? Following the push forward, there is a contraction of longitudinal muscles accompanied by a relaxation of circular ones. This expands the anterior region of the worm against the soil, and the animal is anchored by setae, bristles in the body wall that can be felt by running a finger posteriorly down the worm. Notice that the animal requires a traction-producing surface in order to move effectively. Of the surfaces you provided, which affords the best traction?

3. Follow the waves of muscular activity in forward and backward movement and in anchoring. What modifications must be made in the symmetrical contractions if the worm is to make turning movements?

4. In order to section the nerve cord, try making a small cut in the ventral side of the worm. Notice the effect on movement, and interpret your observations.

Exoskeletons Examine representatives of the phylum Arthropoda, preferably alive. These animals have a rigid exoskeleton, composed principally of the polysaccharide **chitin**, covering most of the body. The exoskeleton is secreted by the underlying epidermis.

The exoskeleton functions to prevent excessive water loss and to protect the internal organs from mechanical injury. In addition, it provides a point of attachment for muscles and forms lever systems between movable parts of the skeleton.

If a crayfish or fiddler crab is available, watch its movement carefully. Rapid locomotion and fine gradations of movement are possible for animals with jointed skeletons. Notice the joints and the manner in which they are manipulated during movement. What are some of the problems that result from having an exoskeleton?

Endoskeletons Vertebrate animals have endoskeletons—internal skeletons composed of bone and/or cartilage, with the muscles and other soft parts of the body outside. The relatively large amount of muscle tissue supportable by an internal skeleton as well as the remarkable internal and external structural adaptability of bone make possible a wide variety of different types of locomotion in vertebrates, despite the serious problems of support inherent in the heavier vertebrate body.

Examine the vertebrate skeletons on demonstration. This outline refers to the human skeleton, but it is broadly applicable to most other vertebrates and reference will often be made to other commonly available skeletons (e.g., frog, cat). You will find it much easier to appreciate skeletal adaptations and to learn the names of a few parts of the skeleton if you try to think about how a particular skeletal structure is organized to carry out its special functions. One of the easiest ways to understand function is to compare as many skeletons as possible in the time available. It will help you to remember parts if you label Figures 13-2 through 13-5 as your study proceeds.

The vertebrate skeleton is arbitrarily divided into **axial** and **appendicular** components. The axial skeleton is the main longitudinal portion, and the appendicular skeleton includes the bones of the appendages and their supportive pectoral and pelvic (shoulder and hip) girdles.

Axial skeleton 1. The skull is composed of many small bones fused together. Note particularly:
   a. The **cranium**, a bony case enclosing the brain and major sense organs (which sense organs?).
   b. The **facial bones**.
   c. The **mandible** (lower jaw), the only bone of the skull not immovably fused to the rest.

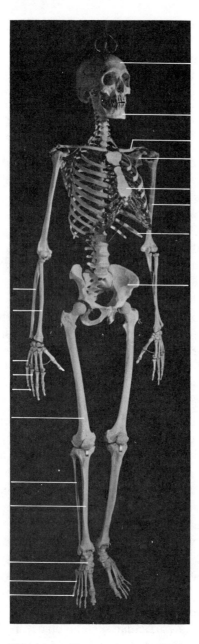

Fig. 13-2. Human skeleton. *(Courtesy Ward's Natural Science Establishment, Inc., Rochester, New York.)*

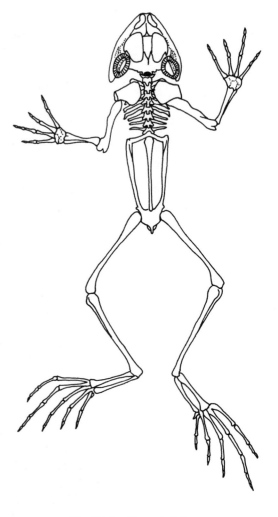

Fig. 13-3. Frog skeleton.

Locate the **hard palate,** the roof of the mouth, in the human skull. Note that it separates the nasal passages from the mouth. You will recall from your study of this region in the rat that the separation is continued posteriorly by the soft palate so that the air passages open into the pharynx far back in the mouth. Such an arrangement makes possible simultaneous breathing and mastication of food (try it). It would be difficult for a mammal to interrupt breathing every time it had to feed, since its metabolic rate is high enough to require regular and constant breathing. Examine the skull of a frog, and notice that the hard (secondary) palate is lacking; the nasal pas-

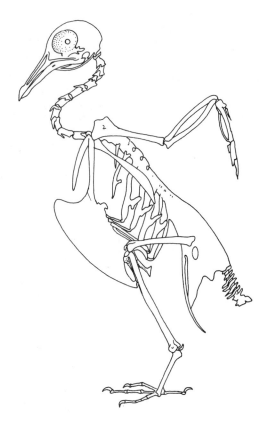

Fig. 13-4. Pigeon skeleton.

sages open near the front of the mouth. Remember that a frog's metabolic rate is lower than that of a mammal and that a frog has alternative sites for gas exchange, whereas a mammal has only its lungs.

2. The **vertebral column** serves for support and for housing the spinal cord and is composed of many vertebrae, separated from one another in life by intervertebral **discs**. The column is customarily divided into five series:

    a. (7) cervical vertebrae (forming the neck region).
    b. (12) thoracic vertebrae (with which the ribs articulate).
    c. (5) lumbar vertebrae (in the abdominal region).
    d. (5) fused sacral vertebrae (enclosed by the pelvic girdle).
    e. Several caudal vertebrae (forming the tail, and fused in man to form the **coccyx**).

The number of vertebrae in each region differs from species to species, and the variation is particularly marked in the caudal vertebrae. The numbers above, in parentheses, refer to human beings.

Notice the stout, dorsal-projecting processes (neural spines) on the vertebrae and how they differ in length. Compare the length of these processes on a human skeleton with those in a cat skeleton. The difference in relative length is particularly marked in the cervical and thoracic regions. Why?

3. Locate the ribs, forming a bony cage that with its associated muscles forms a support for the chest wall. The ribs articulate dorsally with thoracic vertebrae, and some are also attached by cartilage directly or indirectly to the **sternum**. Those ribs without ventral attachment are called floating ribs.

The frog has a small sternum (partly bone and partly cartilage), but it lacks ribs. What functions in mammals are performed by the ribs? How are these functions performed in a frog?

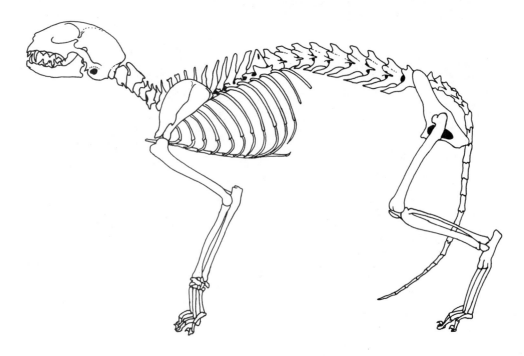

Fig. 13-5. Cat skeleton.

Examine the skeleton of a bird, if available, and note the remarkable relative difference in size between the sternum of a bird and that of a mammal. The large area of the bird sternum provides a surface for the attachment of wing muscles, a function that the mammalian sternum does not have to perform, although the latter does serve as a point of attachment for muscles that support and flex the arm toward the body axis.

Appendicular skeleton

4. The **pectoral girdle,** supporting the forelimbs, is composed of two pairs of bones:
   a. **Clavicles** (collar bones).
   b. **Scapulas** (shoulder bones).
Note the prominent clavicle in humans and the very tiny one in cats (in some cat skeletal preparations, it may be absent). The clavicles provide pivot points in animals that can move their forelimbs in many different directions but interfere with movements of these limbs during running (how?). The clavicles also provide support for animals that hang or lift with the forelimb.

5. The forelimbs are composed of loosely articulated bones, as follows:
   a. **Humerus,** the large long bone of the upper portion of the limb.
   b. **Radius,** the long bone of the lower arm, with a ball-and-socket joint at the elbow, allowing rotational motion.
   c. **Ulna,** the other long bone of the lower arm, with a hinge joint at the elbow, allowing motion in only one plane.
   d. **Carpals,** a group of small bones forming the wrist.
   e. **Metacarpals,** slender bones forming the palm.
   f. **Phalanges,** the bones of the fingers.
Compare the radius and ulna of the human or cat skeleton with those of the frog, in which rotational motion is neither necessary nor desirable.
The position of the ulna makes it useful as a point of attachment for

muslces, but it bears very little weight relative to the radius. In some animals, the ulna is, therefore, reduced in size or partially fused with the radius. If the forelimb of a horse is available, note the fusion of radius and ulna, with only the upper portion of the ulna still clearly defined.

Notice the difference between carpals, metacarpals, and phalanges in cat and human skeletons. The cat stands on its fingers and toes, an arrangement that gives it additional leverage for running.

6. The **pelvic girdle** forms the basal support for the hindlimbs and is composed of three pairs of tightly fused bones.

7. The hindlimbs are composed of a series of loosely articulated bones, as follows:

      a. **Femur,** the long bone of the thigh.

      b. **Tibia,** the larger of the two long bones of the shank.

      c. **Fibula,** the smaller of the two long bones of the shank.

      d. **Tarsals,** a group of small bones forming the ankle.

      e. **Metatarsals,** slender bones of the foot.

      f. **Phalanges,** the bones of the toes.

The fibula is comparable to the ulna in that it bears little weight relative to the tibia and may, therefore, sometimes undergo reduction or fusion with the larger bone.

Bone and Cartilage   Most elements of the endoskeleton are preformed in cartilage and later replaced by bone, but some parts remain permanently cartilaginous. Cartilage and bone are two types of *connective tissue* in which the cellular elements secrete a **matrix** (fibers and a polysaccharide gel in cartilage or calcium salts in bone).

Examine slides of hyaline cartilage and compact bone in cross section. Study the demonstration comparing longitudinal sections through long bones of birds and mammals, if available.

1. In the slide of **hyaline cartilage,** note the clear, noncellular gel matrix (the fibers will not be visible) and the cavities containing cartilage cells. Depending upon the area from which the tissue for your slide was taken, you may also find some fibrous connective tissue and fat associated with the cartilage.

2. Bones have a complex internal structure. In a long bone such as the femur, the outer and stronger parts usually consist of compact bone, and the inner region, often hollow and traversed by bars of bony material, is spongy bone. Crossbars in the spongy region support the whole bone in much the same manner that the superstructure of an airplane is supported by steel struts. The exact organization of spongy bone depends on the mechanical stresses to which the bone is subject. Marrow, which fills out the spaces in spongy bone, serves as a region of blood-cell manufacture and as an area of fat storage.

Birds, in which skeletal lightness is advantageous, lack spongy bone and marrow although crossbars are present to provide strength. The interior of many bird bones contains air sacs which contribute to lightness.

3. Examine a cross section of compact bone. Notice the many **Haversian systems,** units of concentric deposition of the mineral-salt matrix of bone. The lumen of each Haversian system, called a Haversian canal, carries blood

Fig. 13-6. Photomicrograph of a cross section of a bone, showing Haversian systems. *(Courtesy Thomas Eisner, Cornell University.)*

vessels and nerves. Outside the lumen, the matrix of the bone is deposited in concentric layers, perforated by dark structures (regions where the bone cells lived) and by tiny radially oriented channels called canaliculi. These channels connect the bone cells to one another and to the Haversian canal. In a mature Haversian system, the bone cells are isolated from one another, and the exchange of materials with the vascular system occurs by means of diffusion from individual cells to the canal through fluid in the canaliculi.

The complicated internal structure of compact bone is the result of changes that occur as the bone develops and as different stresses are placed upon it. The mineral salts (calcium phosphate and calcium carbonate) of which the bone matrix is composed are constantly undergoing resorption in some areas to form small, tubular channels. New matrix material is deposited in concentric layers on the walls of these channels. This continual reformation and reworking of bone accounts for the appearance in cross section of the Haversian systems, each of which was once an area of resorption.

Label the photomicrograph of a cross section of compact bone (Figure 13-6). Why would it be adaptive for compact bone to be reworked and organized in this complicated manner? In what parts of the body would you expect to find bony structures that did *not* show the development of Haversian systems?

THE ARRANGEMENT AND ACTION OF SKELETAL MUSCLE

Muscle Arrangement   Working in groups as directed by your instructor, dissect and study the arrangement of muscles in the hind leg of a frog.

Using a doubly pithed frog, cut all the way around the loose skin next to the hip at the base of one of the rear legs. Grasp the skin firmly, and pull it back, exposing the muscles of the thigh and shank.

Locate the **gastrocnemius,** a large muscle on the back of the shank. Observe the white connective tissue (fascia) surrounding the fleshy belly of the muscle. The fascia continues beyond the muscle at either extremity to form strong white bands, the **tendons,** that attach the muscle to bones (Figure 13-7).

When a muscle contracts, one of its extremities (usually the proximal one) remains essentially stationary. This point of attachment is called the **origin** of the muscle. The movable attachment is called the muscle's **insertion.** The gastrocnemius has two proximal origins: the larger one is on the posterior portion of the distal end of the femur, near the knee; the smaller

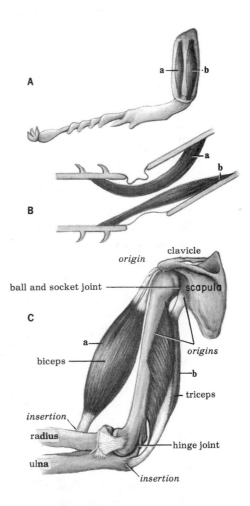

Fig. 13-7. Mechanical arrangement of muscle and skeleton in a human arm and in insect legs.

forms a small tendon, attached to a larger, sheetlike tendon covering much of the anterior portion of the knee. The large tendon attaches directly to the femur and is actually the insertion of one of the large thigh muscles—the triceps. At its distal end, the gastrocnemius inserts on the ankle and foot by way of the long **Achilles tendon,** which extends over the heel.

Muscle Action   The type of action resulting from contraction of a particular muscle depends upon the exact position of the muscle's origin and insertion, the type of joint between them, and the action of other muscles—both **antagonists,** arranged to oppose the action of the muscle in question, and **synergists,** which guide and limit the principal movement, thereby making possible finer gradations of motion.

There are several terms commonly applied to describe the action of muscles:

1. Flexor—causes jointed parts to move toward each other.
2. Extensor—causes jointed parts to move away from each other.
3. Adductor—causes movement of a part toward a vertical plane running through the longitudinal mid-line of the body.
4. Abductor—causes movement of a part away from a vertical plane running through the longitudinal mid-line of the body.

Hold the thigh rigid with one hand, and grasp the belly of the gastrocnemius with the other. Gently pull the gastrocnemius toward its origin. What is the action of the muscle under these conditions? Next hold the shank rigid, and again pull the gastrocnemius toward its origin. Now what is the action of the gastrocnemius? Decide by examination which muscles, acting as antagonists, determine for any given instance which of these two major actions the gastrocnemius performs.

Locate the **triceps femoris,** a very large muscle on the anterior surface of the thigh (it is best seen in dorsal view). This muscle has two origins on the pelvic girdle and inserts on the femur by way of the broad white tendon already seen passing over the anterior surface of the knee. Hold the thigh firmly, and pull on the triceps. Is this muscle an extensor or a flexor of the shank?

The **biceps femoris** is a narrow muscle best seen in dorsal view just posterior to the triceps. Its origin is on the pelvic girdle. Where does it insert? What is its action?

MUSCLE PHYSIOLOGY

One way to study the characteristics of muscle contraction is to stimulate a muscle and watch it. This procedure will tell you that muscles, like nerves, respond to many kinds of stimulation (touch, chemicals, heat, electricity, drying, etc.), and it can give you a crude idea of the relationship between the amount of stimulation and the extent to which the muscle responds.

Much more precise information can be obtained by using the kymograph apparatus, one variation of which is illustrated in Figure 13-8. Using this arrangement, an electrically stimulated muscle is attached to a lever that

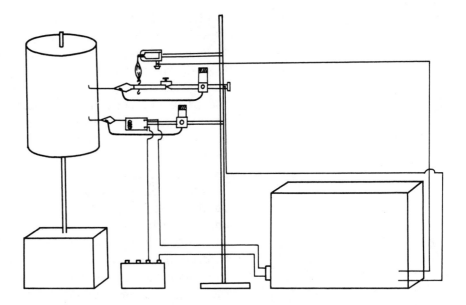

Fig. 13-8. Kymograph apparatus, for studying muscle contraction. Fill in and label the controls present on the instrument used in your laboratory.

amplifies the contraction and records it by writing on paper attached to a moving drum. Simultaneously, an electrical device, the signal marker, records the application of the stimulus so that the relationship between the response of the muscle and the application of the stimulus can be ascertained. This arrangement has many advantages over *in situ* stimulation of muscles. For one thing, very small responses can be detected because of amplification through the lever system. In addition, the moving drum allows a detailed examination of the different phases of contraction and relaxation. The application and measurement of stimuli are much more readily controlled than *in situ*.

Electrical stimuli are used because they are easily applied and controlled for intensity, frequency, and duration; because they do not damage the muscle; and because they are most like the muscle's natural stimulation in the body.

Your instructor may have you perform these experiments in groups, or the experiments may be done as a demonstration. In either case, learn the parts and operation of the apparatus *before* the experiments are done, and find out if and how your apparatus differs from that illustrated. As you identify the parts, label them in Figure 13-8, adding any part not already in the diagram.

Parts of the Apparatus

1. The **stimulator** is a device used to generate electrical stimuli of controlled duration, frequency, and intensity (voltage). Check the following controls, adding to Figure 13-8 those which are present on your stimulator:

    a. Signal marker switch—set at the "On" position. This is a device that gives a "click" when your stimulator delivers a shock to the muscle. You, therefore, have an audible and a visual (see 4, below) check on the delivery of stimuli.

b. Frequency dial—set at the lowest frequency. This adjustment controls the number of times per second that stimuli are delivered to the muscle. Most stimulators have a "Multiply By" dial adjacent to the frequency control dial, which usually permits an increase in the frequency by multiples of ten. The total frequency in any instance is the value shown on the "Multiply By" dial times that shown on the frequency dial. Set the "Multiply By" dial at 0.1 or its lowest setting.

c. Duration dial—set as specified by your instructor. This adjustment controls the length of time over which the stimulus is delivered, and the setting is usually very small. Set the "Multiply By" dial, adjacent to the duration dial, at 0.1 or its lowest setting.

d. Voltage dial—set at the lowest value. This adjustment permits you to control the intensity of the stimulus, measured in volts. As for the frequency and duration adjustments, total intensity is calculated by multiplying the value shown on the "Multiply By" dial by the value shown on the voltage dial. Set the "Multiply By" dial at its lowest value or 0.1.

e. Mode switch—set on "Off." This switch permits the operator to deliver single stimuli at his or her own rate (usually the "down" position of the switch, but check the label carefully) or to deliver continuous or "repeat" stimuli at a selected frequency.

f. Stimulus selector—set on "Regular."

g. Output—set on "Mono."

h. Polarity—set on "Normal."

i. Power switch—set on "On."

Check to see that the stimulator is properly plugged in. The wiring leaving the stimulator and running to the rest of the apparatus should not be changed from one laboratory section to the next. If you find free-flying wires, consult your instructor before proceeding.

2. The **kymograph** consists of an electrical or spring-driven motor that powers a revolving drum mounted on a vertical post. Your instructor will explain how to control its speed and will tell you how to mount paper upon the drum. The drum is covered with a glossy paper, which may be plain if the writing is to be done in ink or may be smoked with carbon particles, if the writing is to be done by scraping off the carbon. To give a longer paper, your apparatus may have an additional vertical post and drum. Mount the drums as high as possible on the posts, and follow your instructor's directions for further handling.

3. The **femur clamp** holds the bone to which the experimental muscle is attached.

4. The **writing lever** is attached to the bottom of the muscle and serves to record its movements on the kymograph paper.

5. The **signal marker** is a device that writes on the kymograph paper to indicate the points at which stimuli are applied to the muscle.

The femur clamp, writing lever, and signal marker should all be arranged so that the muscle hangs vertically after it has been attached to the apparatus, and so that the two writing points contact the drum with approximately equal pressure. Adjust the position of the levers *after* you have attached the muscle. If pressure on the drum is too great, the levers will lag in their movements due to friction against the paper; if the pressure is too slight, the lever will come off the drum at some points in its movement. Either situation will produce an uneven record. If you are using ink-writing apparatus and it is not operating properly, call your instructor.

Read through *all* the experiments before you prepare the muscle.

Preparation of the muscle (Figure 13-9)

1. Cut around the skin at the base of the intact leg, just as you did in examining the muscles earlier (Figure 13-9A).

2. Skin the leg as if you were removing a glove from a hand (Figure 13-9B).

3. Cut off all muscle from the thigh bone (femur) *except* the gastrocnemius origin, which should be left intact.

4. Insert the tip of the scissors between the calf muscle and the shank bone (tibiofibula), and carefully separate the muscle from the bone. Then insert the scissors tip between the Achilles tendon and the heel. Cut the tendon, leaving as much as possible of it attached to the gastrocnemius. Completely separate the gastrocnemius with its tendon from the remainder of the lower leg, up to the knee. Cut away all muscle and bone *except* the gastrocnemius, the Achilles tendon, and the femur to which the gastrocnemius is attached. The resulting preparation should look like Figure 13-9C.

5. Moisten the muscle with frog Ringer's solution. Insert the femur into the femur clamp, and fix it firmly therein. Carefully hook a small S-shaped pin through the Achilles tendon, and hook it to the writing lever. Put a small amount of tension on the muscle by moving the femur clamp on its vertical post. Adjust the position of the levers relative to the drum (as explained in "Parts of the Apparatus," number 5). The writing levers should be horizontal and should have some tension on them so that when the muscle contracts slightly the lever will record the movement.

6. One person on each team should be responsible for keeping the muscle moist with frog Ringer's solution throughout the experiments. The solution may be dropped over the muscle with a medicine dropper, but be careful *not* to spill any on the apparatus.

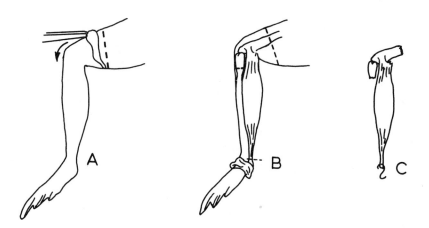

Fig. 13-9. Steps in the preparation of a frog muscle.

Experiments

Threshold

1. For any given stimulus duration and intensity, there is a minimum voltage—the **threshold voltage**—below which the muscle will not respond. With the duration and frequency set on their lowest values (or otherwise, as specified by your instructor), start the voltage at its lowest value, and stimulate the muscle with single shocks a few seconds apart. Increase the voltage slightly, moving the paper about 1 cm each time, until you see a response. Record this voltage—along with the frequency and the set-duration value—as the threshold voltage. (*Note:* If no response occurs, even if you have advanced the voltage dial to its *highest* value, decrease the voltage dial to its *lowest* value, and turn the "Multiply By" dial up one factor—usually from 0.1 to 1.0. Then begin turning up the voltage dial again and testing with single shocks until a response occurs.)

Maximal Stimulus

2. After reaching the threshold, continue to increase the voltage, stepwise, turning the drum each time. Note that as the voltage is increased, the extent of the response increases as well. This is explained by the fact that the muscle contains many fibers, not all of which have the same threshold. As the stimulus intensity is increased, more and more fibers contract and the muscle shows a larger and larger movement. Eventually you will reach a **maximal** value, beyond which further increase in voltage will not produce an increase in contraction. Any stimulus applied after the maximal stimulus is called a **supramaximal** stimulus.

Individual muscle fibers respond like nerve fibers—in an all-or-none fashion. The muscle as a whole, however, can respond incrementally to increasing stimulus intensity because the individual fibers have different thresholds.

Single muscle twitch

3. Arrange the muscle for a single contraction and relaxation—a single "twitch"—with the drum turning at full speed. Label the curve with contraction period and relaxation period. The **latent period** of the muscle, that interval between the arrival of the stimulus and the event of contraction, is too small to be measured without additional special equipment. Refer to your text for details.

Summation of response, and tetanus

4. Set the voltage at the level discovered as maximal stimulus for your muscle, and set the kymograph speed on high, without yet turning it on. Set the mode switch to repeat, and with the drum turning, gradually increase the stimulus frequency from 2 to 20 per second. You will notice during the increase in frequency that the muscle does not fully relax after each contraction, but is hit by another stimulus during its relaxation period and begins to contract again, stronger than before. As the frequency increases, this effect, called **summation** of response, becomes more and more marked. How might you explain this greater contraction than that elicited by a maximal single stimulus? Finally there is *no* relaxation at all, but only a smooth curve. This condition is called **tetanus**. Normal contraction of skeletal muscles is tetanic contraction, brought about by the delivery of very frequent stimuli from the central nervous system. If skeletal muscles did not contract in this manner, muscular movements would be jerky and uncoordinated.

Fatigue

5. With a frequency setting just below that needed to obtain summation and with the voltage setting required to get maximal response from the muscle, allow the apparatus to run slowly until the muscle fatigues and begins to relax, even though stimuli continue to impinge on it. What does the fatigue curve look like? Why does the contraction not continue to be full? What biochemical events take place during fatigue?

6. When you have finished the exercises, set the stimulator dials at the lowest levels, turn off the power, unplug the stimulator, clean up the apparatus and your desk area, and turn in your fully labeled record to the instructor.

THE MOLECULAR BASIS OF MUSCLE CONTRACTION

While we know much about the mechanics and physiology of muscle activity, the mechanism by which a muscle fiber shortens is still imperfectly understood. The most widely accepted hypothesis, which accounts for the difference in ultrastructural appearance of relaxed and contracted muscle cells, shows that contraction involves two muscle proteins: actin and myosin. The molecules of these proteins are believed to slide just past one another during the contraction, from an extended position to an apposed, shortened position. The sliding will not occur without ATP and certain mineral ions.

Observing the influence of ATP

In this exercise, you will have an opportunity to observe directly the influence of ATP and of the mineral ions on the contraction process, and to obtain a quantitative measure of contraction. The material used in these experiments is psoas muscle, one of the skeletal muscles lying along the backbone. It was removed from a freshly dead rabbit, tied on a stick at approximately its natural body tension, placed in a solution of 50% glycerol, and stored in a freezer. The glycerol removes almost everything except the contractile proteins, but the muscle retains its striated appearance and ability to contract.

Each group of two to four students will be provided with a 2-cm section of psoas muscle in a dish of cold glycerol. Using glass needles or *clean* stainless-steel forceps, one member of the group should separate the muscle into bundles, one for each student. Each bundle should be placed in a Syracuse dish of glycerol solution. From this point on, each student should follow the procedure on his or her own.

1. Using 7–10X magnification if possible, and using glass needles or clean stainless-steel forceps (preferably the latter), gently tease the muscle into very thin groups of myofibers (the muscle cells). Single cells or thin groups are best; strands thicker than a silk thread will curl up when they contract and must not be used.

2. Mount one of the strands in a drop of glycerol on a slide, with a cover slip. A Pasteur pipette or dropper is best for transferring the fiber. Examine under low and high powers. Note the striations, and compare your observations with photograph 2 in Figure 13-10. Notice that each cell has many nuclei.

3. Transfer three or more of the thinnest strands to a drop of glycerol on a second slide. The strands should be the same length, and a cover slip should *not* be added. Using a dissecting microscope and a millimeter ruler

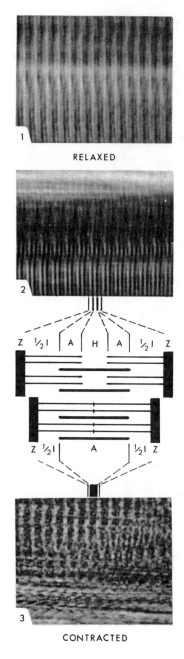

RELAXED

Z | ½I | A | H | A | ½I | Z

Z | ½I | A | ½I | Z

CONTRACTED

Fig. 13-10. Glycerinated muscle fibers, phase contrast 210X. 1 and 2: Relaxed muscle; 3: contracted muscle. The drawing between 2 and 3 shows the banding difference between the relaxed and contracted muscle. *(Courtesy Carolina Biological Supply Company.)*

held beneath the slide, measure the length of the fibers. If there is more than a small drop of glycerol on the slide, soak up the excess on a piece of lens paper held at the edge of the glycerol farthest from the fibers.

4. Flood the fibers with several drops of a solution containing both ATP and the required mineral ions. Observe the reaction of the fibers, and, after thirty seconds or more, remeasure the length of the fibers. Calculate the degree of contraction.

5. Repeat the experiment twice, using clean slides and pipettes, and being especially sure to use *clean* teasing needles or forceps. In one of the repetitions, use ATP alone; in the other, add metal ions alone. Record and explain the results. (*Note:* If you contaminate either of the solutions with the other, however indirectly, you will obtain ambiguous results. If your re-

sults are ambiguous or otherwise unexpected, compare your procedure with that used by your neighbors and try to explain your results.)

6. Observe one of the contracted fibers under low and high powers of a compound microscope, and look for differences in appearance between the muscle in a contracted and in a relaxed state. If the differences are not apparent in your preparation, note them in Figure 13-10 along with a diagram that relates the banding pattern seen in the muscle to the hypothetical sliding of filaments.

It appears that the shortening of a muscle fiber is due to the shortening of its constituent sarcomeres. This, in turn, results from a progressional sliding of the thin filaments inward between the thicker filaments, pulling the dense "Z" membranes nearer together. The free ends of opposing thin filaments eventually may meet one another at the mid-line of the sarcomere. Striations in contracted glycerinated muscle are thus distinguished from those in relaxed muscle by an obliteration of the "H" discs and a decrease in length of the "I" bands. Figure 13-11 compares relaxed and contracted muscle in electron micrographs and should make these differences clear.

Fig. 13-11. Electron micrograph of a skeletal muscle from a rabbit. The myofibrils run diagonally across the micrograph from lower left to upper right; each looks like a ribbon crossed by alternating light and dark bands. The wide light bands are I bands; there is a narrow Z line in the middle of each. The wide dark bands are A bands, each with a lighter H zone across the middle. *(Courtesy H. E. Huxley, Cambridge University.)*

QUESTIONS FOR FURTHER THOUGHT

1. Write a concise but thorough description of the physiology of muscle contraction at the nonmolecular level. Be sure to include an explanation of each of the following: all or none, threshold, maximal stimulus, summation of response, tetanus, latent period, fatigue.

2. Explain the theory of the molecular physiology of skeletal muscle contraction proposed by H. E. Huxley, and give some of the evidence on which this theory is based.

3. Explain the skeletal and muscular arrangements and functions that make it possible to use the same body part to perform many different motions or to perform the same motion at many different speeds. Include mention of the following: origin, insertion, antagonist, synergist, joint, etc.

4. What functions, apart from support, does the endoskeleton perform?

5. In a single sentence, explain the operation of the hydrostatic skeleton in an earthworm. How would the repertoire of movements differ if the body were not segmented?

TOPIC 14 BEHAVIOR

Biological Science, 4th ed.
Chap. 15, pp. 378–402; Chap. 21, pp. 553–580
Elements of Biological Science, 3rd ed.
Chap. 14, pp. 236–248, 249–251; Chap. 20, pp. 346–374

One of the basic characteristics of life is its capacity for response to environmental change. An organism reacts to a specific change in its environment, a **stimulus,** by carrying out some activity that we call a **response.** The repertoire of responses that characterizes an organism is called its **behavior.** Living organisms have evolved a remarkable diversity of behaviors, which, for convenience of discussion, we may group into two categories: learned behavior, that array of responses acquired by an organism in the course of its experience, and inherited or innate behavior, which is a part of the organism's genetic endowment. In this laboratory, we shall study a few examples of inherited behavior.

Like the nervous and effector systems from which it is molded, behavior is subject to evolutionary adaptation. As you watch different behaviors today, try to think about them as adaptive mechanisms for the organisms in question and ask yourself how they may be useful to the organism in its natural situation. Think also about how the stimuli you are using may differ from those to which the organism is normally subject. Discussion with your neighbors and comparison of results are encouraged.

Several shorthand terms will be helpful to you in describing the behaviors you see in this laboratory and also in thinking about their adaptive value. Locomotor behavior, that involving movement of an organism from one place to another, is described by two terms: **taxis** and **kinesis.** A taxis is an oriented movement toward or away from the stimulus that brought about the movement. A kinesis, on the other hand, is random movement caused by a stimulus but not necessarily oriented by it. Some insects, for example, select a dark habitat in preference to a lighted one. If we can demonstrate that an animal actually avoids light by moving directly away from it, we can describe its behavior as a taxis (in this case, a phototaxis—i.e., one caused by light). It is possible, however, that light, or the "absense of darkness," may merely initiate random movements which eventually chance to carry the animal into the darkness, where the movement "turns off." This sort of behavior would be called a kinesis. The terms "taxis" and "kinesis" are most often used to describe animal behavior, but they apply equally well to any organism that shows locomotion—for example, the motile algae.

Two other terms we shall find useful are **nastic movement** and **tropism;** we shall use them to describe nonlocomotor behavioral movements in plants. A nastic movement is a change in the posture of a plant or plant part, and it is the same no matter what the stimulus direction. It does not orient the plant toward or away from the stimulus, as could the locomotor behaviors we discussed above. A tropism, in sharp contrast, is a growth curvature, a postural change oriented by the direction of the stimulus.

We may add a prefix, as shown in the following list, to any of the terms discussed above to describe the nature of the stimulus. We may add the word "positive" or "negative" to describe the direction of the movement relative to the stimulus. For example, the bending of a sunflower toward the sun would be called a **positive phototropism**.

photo-	reaction to light
thermo-	reaction to temperature
geo-	reaction to gravity
hydro-	reaction to water
hygro-	reaction to moisture or humidity
thigmo-	reaction to contact
chemo-	reaction to chemicals
galvano-	reaction to electricity
opto-	reaction to change in the visual field
seismo-	reaction to shock

Do the following experiments in groups, as directed by your instructor. Your group may be asked to give a report on one of the experiments. In the report, you are expected to pool the results obtained by the rest of the class and to summarize information regarding the adaptive value of the behavior you discuss.

PLANT MOVEMENTS

Mimosa

Obtain a potted specimen of the sensitive plant *Mimosa*, and, *without touching the leaves,* move it carefully to your desk.

1. Stimulate one of the end leaflets by gently touching it. If nothing happens, stimulate it again by touching it with a little more force. What is the plant's response? How would you describe this response, using the terminology in the introduction? What benefit might the plant derive from this behavior? Note the approximate time when the response was given.

The movement you saw was due to a rapid change in the turgor of cells at the base of the leaflet. Can you locate in this region any structure that might help to transmit the pressure of your touch to these sensitive cells?

2. Time the recovery of the leaflets to their normal posture. Design and carry out an experiment to test the effect of increased severity of stimulus, or try using other stimuli. You might wish to try to find out whether the plant is particularly sensitive to touch in certain regions of the leaflets. While you are waiting for the plant to recover, go on to the next part.

Photonasty in Oxalis

Obtain an *Oxalis* plant (Figure 14-1) that has been kept in sunlight, and examine the leaves and stalk. In the space provided (Figure 14-2), sketch the angle of the leaves relative to the stalk.

1. Place a cardboard box over the plant, and leave the plant alone for about an hour. In the meantime, go on to other parts of the laboratory. At the end of an hour, remove the box just long enough (less than a minute) to glance at the plant so that you can again sketch the angle of the leaves rela-

Fig. 14-1. Photograph of *Oxalis*. On the right, notice the *Mimosa*.

tive to the stalk. Immediately replace the cardboard box, and sketch the angle in Figure 14-3.

2. With the box still covering the plant, design and carry out an experiment to prove that this behavior is a nastic movement (that its direction is not determined by the stimulus).

Fig. 14-2. The angle of *Oxalis* leaves before the experiment.

Fig. 14-3. The angle of *Oxalis* leaves after the experiment.

Fig. 14-4. *Euglena,* diagram of structures seen under oil immersion.

Phototaxis in an algal protist You will be supplied with a petri dish containing *Euglena,* an algal protist. Figure 14-4 diagrams the structure of *Euglena.* Make a wet mount for comparison.

1. Stir the water in the dish gently until the organisms are equally dispersed throughout the container. Cover the dish completely with aluminum foil, making certain that no light can enter. After half an hour, remove the foil, and record the distribution of the organisms in a sketch.

2. Cut a small circle (5-mm diameter) in the aluminum-foil cover, and put the cover back on the container. Illuminate the *Euglena* through this hole, using a microscope lamp or other source positioned to deliver a vertical beam of light into the container, and set up far enough away so that the water will not become heated. After half an hour, remove the foil cover, and record the distribution of the organisms. What would you call this response? How would you design an experiment or observations to determine whether these were taxes or kineses?

3. Design and carry out an experiment to demonstrate negative taxis. How could you design an experiment that would demonstrate negative *photo*taxis? Eliminate any possible effects of a temperature stimulus.

Demonstration of Plant Tropisms Examine the demonstration of plant tropisms. Each setup will be labeled with an explanatory card. Distinguish between positive and negative phototropism and geotropism.

If there is a film of plant tropisms, be certain to see it sometime during the period. List the kinds of tropism seen in the film.

SLIME-MOLD BEHAVIOR

During this part of the laboratory, you will have the opportunity to work with an unusual organism, the slime mold *Physarum polycephalum.* One important purpose of this exercise is to illustrate, by means of relatively simple observations and experiments, how questions may arise and be answered or how more questions may be generated in the process.

Below you will find suggested observations and experiments which you may wish to do, according to your interest and the time available. You will not be able to do them all, and you should feel free to try any experiments that occur to you, whether or not they are included in the list. Keep full notes, and in the next laboratory period turn in a report (as directed by your instructor) of your experiments and observations.

You will be provided with cultures of the **plasmodium** stage of *Physarum*. Figure 14-5 shows the life cycle of the mold. To study the organism, remove the cover of the petri dish; but do not leave it off for long periods when you are not observing the organism, or the culture will dry out. If you wish to do experiments that will damage or kill the cultures (such experiments in the following list are indicated by an asterisk), make certain you check with your instructor before you proceed. In any case, save the damaging experiments until all of the others you wish to do have been completed.

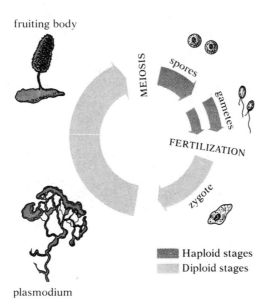

fruiting body

MEIOSIS

spores

gametes

FERTILIZATION

zygote

■ Haploid stages
▨ Diploid stages

plasmodium

Fig. 14-5. Life cycle of a true slime mold. See the text for a description.

Observing Physarum

1. In nature, slime molds are found on the forest floor among rotting leaf mold and other decaying vegetation. They slowly move about, absorbing bacteria and decaying matter. How would the oat flakes on the agar contribute to nourishing the mold? If the culture were sterile, would the mold be able to feed?

2. What color is your specimen? How would you design an experiment to determine whether color is a heritable trait or caused by some other factor such as type of food ingested?

3. Examine your culture without the aid of a microscope. Can you see any evidence that the culture has recently undergone changes in shape, size, or position on the agar?

4. Examine the thin white tracks on the agar. Explain their presence. How can these explanations be tested?

5. Using different methods of microscopy and lighting, observe the regions in your mold where cytoplasmic streaming is taking place. The best regions are those of relatively new growth, where there are many small interlaced branches. Determine the diameter of the field in your microscope.

From this information, estimate the rate of streaming in millimeters per minute.

6. Is the rate of streaming consistent within a single branch? Is it the same in different branches on the same specimen?

7. If you observe closely, you will notice that the direction of streaming is constantly changing. Can you detect rhythm or regularity in these oscillations? Does the streaming occur in both directions for exactly the same length of time?

8. Is there any consistency in rate of streaming or in rate of oscillation between your specimen and others in the class? (If you discover that the answer is "yes" after observing consistent regularities in data from other students, can you explain the consistency?)

9. In regions where streaming is taking place, do *all* parts of the plasmodium move? Describe the physical arrangement of sol and gel protoplasm in a single streaming branch.

10. Can you detect any structures that might be responsible for the movement? Where do streaming particles originate, and where do they go?

11. Can the mold stream *upward* on a vertically oriented piece of agar?

12. Devise a means of discovering how rapidly the mold as a whole moves across the agar.

13. Do you think the mold can sense the presence of food at a distance, or is contact with food made entirely by chance? Design and carry out an experiment to test your hypothesis.

14. Devise an experiment to discover whether the pattern of peripheral branching differs in fed and unfed specimens. Explain the results.

*15. With a dissecting needle, make a delicate puncture into a region where active streaming has been observed. If possible, make the puncture while you are observing the mold under the microscope. What is the immediate response? Record the response over a five- to ten-minute period. Is the puncture apparently repaired during this time? Does the streaming return to normal?

*16. With the edge of a clean razor blade, make a delicate, narrow cut across the edge of a branch. Examine the cut at intervals. Do the severed regions rejoin?

17. Test the mold's response to cold by placing the covered petri dish on crushed ice for ten minutes. Describe the reaction. Can the mold recover from its response to cold? (Be careful *not* to get ice into the culture dish.)

*18. Heat a glass rod in an alcohol lamp, and quickly bring the rod close to (but not touching) a streaming region of the plasmodium. What is the response?

19. Your slime mold has been cultured in the dark, where these organisms usually grow. Cooperating with others at your table, set up a *properly controlled* experiment to determine the mold's response to window light during your laboratory period.

ANIMAL MOVEMENTS

Reactions to moisture and to light

We will use animals known as **isopods**. These terrestrial crustaceans are sometimes called pill bugs, sow bugs, or volkswagens. While most crustaceans are aquatic, such as lobsters, crayfish, shrimp, the isopods have also established themselves on land. Their survival on land depends on a very moist habitat, and terrestrial isopods are usually found only in very moist places such as under stones, rotting logs, leaf mold, etc.

1. Prepare two glass bowls by putting a 3-inch square of moistened (*not* soaking) paper towel into one and an identical unmoistened square into the other. Obtain ten isopods from the stock container, and introduce five of them into each bowl. Cover each bowl with a box to exclude the possibility of light acting as a stimulus to activity.

After ten minutes, remove the boxes, and *immediately* observe the degree of activity in each bowl. How would you describe the rate of locomotion in each bowl? Is there any directedness to movement in either of the bowls? Using the terms previously introduced, how would you describe this reaction? Can you tell, from this experiment, whether the stimulus is presence or absence of water? Can you be certain that the stimulus has anything to do with direct sensing of liquid water? Could it be a response to presence or absence of *moisture*? How could you set up experiments to test the answers to these questions?

2. Obtain a piece of waxed paper and a pan from the supply desk. Line the bottom of the pan with two pieces of paper toweling, separated in the middle and overlapped by a 2-inch strip of waxed paper (Figure 14-6). Secure the waxed paper with tape. The paper should fit the pan snugly. With distilled water, moisten one of the pieces of paper towel (*not* dripping wet), and leave the other dry. Introduce ten isopods to the waxed-paper strip, and cover the pan with aluminum foil. After ten minutes, remove the foil, and observe the position and activity of the isopods. How many are on the wet paper? How many are on the dry paper? What is the state of activity of the isopods on the wet paper? On the dry paper? Repeat the experiment several times. What conclusions can be drawn?

Fig. 14-6. Isopod chamber with waxed paper dividing wet and dry paper toweling.

3. Design and carry out some experiments using the pan, paper toweling, waxed paper, and foil to show the effect of light alone, or of light and moisture together, as stimuli. When you finish, return all animals to the stock container at the supply desk.

In interpreting these experiments, it is important to recognize the existence of problems in experimental design that may greatly affect the outcome of your experimental procedures. One such problem is *where* the animals are introduced in the pan. If you happen to introduce more of them toward the moistened area, more will naturally be counted there; and it is possible that they might never have been counted there if the introduction of animals had been altogether random. The use of the waxed-paper strip is a step toward the attainment of random introduction, but there are many better methods of solving the problem. Discuss the experiment with the other students in your group, and be prepared to suggest other ways in which the problem of attaining randomness in the introduction of animals could be solved. The simpler your solutions, the better. Your instructor may be able to provide you with some equipment to carry out your improved experiments, if there is time.

Reaction to temperature

Obtain a piece of glass tubing (bent into a 90° angle at each end), two wooden stoppers for the glass tube, two beakers, and a thermometer.

1. Fill the tube with distilled water, and add 1 or 2 drops of a rich *Paramecium* culture from the supply desk. Stopper the ends of the tube.

2. Fill one small beaker with ice and water, and another with water at about 36°C from the hot-water tap. Arrange the beakers and glass tube as in Figure 14-7, placing each end of the tube in a different beaker. After fifteen minutes, record the temperature of the water in each beaker, and, under the dissecting microscope, observe the position of the organisms. Are they distributed uniformly throughout the tube, or have they become concentrated in a specific region? Can you determine whether the behavior is a taxis or a kinesis?

3. Repeat the experiment, using one beaker filled with water at 20°C and the other filled with water at 55°C. Record the temperature of the two beakers at the beginning and at the end of the experiment. After fifteen minutes, observe and record the distribution of the organisms, and explain your results.

Assuming that the temperature within each of the tube ends is the same as that in the beaker in which it is immersed, and assuming that a uniform gradient of temperature has been established within the tube as a whole (probably *not* a valid assumption), prepare a diagram of the tube showing the degree of temperature change per unit of length. Observing the region of the tube where most of the organisms have congregated, determine the optimum temperature for *Paramecium.*

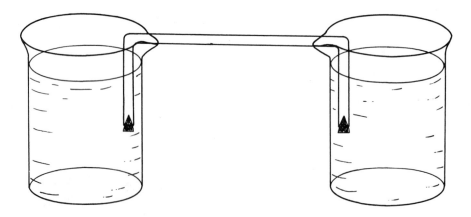

Fig. 14-7. Temperature-gradient setup.

Reaction to gravity

Obtain a turtle from the stock container. Quietly hold it parallel to and about a foot away from the table, until the animal extends its head. Slowly lower the posterior part of the turtle. How does the head move? Then slowly raise the posterior part of the turtle, and note the accompanying head movement.

Repeat the experiment with increased speed, and note the response. Try tilting the animal around its longitudinal axis. What is the result? What sensory devices mediate these responses? Where are they located?

Reaction to change in the visual field

Using the kymograph drum and motor, you will change the visual environment of a fish. Figure 14-8 shows the setup. A cylinder of paper, striped on the inside with tape, has been attached to the top of the drum. When the drum is turned on, the paper and its stripes move with the drum. Also notice that a beaker is hung from a ring stand down into the paper cylinder.

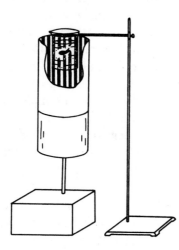

Fig. 14-8. Optomotor apparatus—using a kymograph to change an animal's visual environment.

1. Remove the beaker. Fill it half full with pond water, and add a small fish. Return the beaker to its position on the ring stand, and slowly turn the kymograph to full speed. What does the fish do? How would you describe this reaction? Of what benefit would this response be to the fish in its natural environment?

2. Repeat the experiment, using the turtle. How do you explain the results?

Look up the phylum or division and class of each of the organisms you have studied in this period, and enter each of them, if they are not already included, in the appropriate space in Appendix 2.

CHEMORECEPTION AND BEHAVIOR IN THE BLOWFLY

TOPIC **15**

Dethier, V. G., 1962. *To Know a Fly.*
Holden-Day, San Francisco
Dethier, V. G., 1976. *The Hungry Fly.*
Harvard University Press, Cambridge, Mass.

Human beings depend very little on their chemical senses for survival, but they are unusual in this respect. For most animals, scent is among the most crucial of the senses, and is involved in a remarkably complex repertoire of activities. Natural selection has exploited the potential use of chemical substances as information transmitters and the properties of sensory cells as chemoreceptors in a variety of different ways. Chemicals are widely used to relay information relating to the recognition of territory, the locating of potential mates, and many other types of activity between members of the same species. Nor is chemical communication confined to intraspecific interactions; many animals use chemicals as defense mechanisms, in locating potential enemies, or in locating specific habitats or foods. Insects are acutely able to discriminate among odors, which not only makes possible intraspecific communication and even "tribe" identification within the same species, but has also been seized upon by evolution in the adaptations of many flowering plants. Many such plants utilize odors to attract pollinating insects; without the insects, the plants would often be incapable of sexual reproduction. One European orchid species even mimics the odor of a particular species of female bee in order to attract males of the same species, which pollinate the blossom when they alight on it and attempt to mate with it.

Chemicals are very important as environmental cues that trigger specific behaviors, particularly in animals with relatively stereotyped behavior patterns. In this laboratory, we shall explore a feeding response by chemoreceptors to the presence of sugar, one of many chemically controlled behaviors in the common blowfly *Sarcophaga bullata.* From this study, you will gain some insights into insect behavior and will learn about the nature of the chemoreceptors under consideration. The study is also a good example of how a relatively simple analysis of behavior patterns is carried out.

Before beginning any of the experiments, read the directions carefully, and divide the responsibilities among the members of your group. Prior to each experiment, discuss and clarify the hypotheses upon which the experiment is based (often purposely not clarified in the directions), and consider what each experimental approach can show. Understand the significance of the controls for each experiment. Prepare reports, as directed by your instructor.

HANDLING THE FLIES

Sarcophaga bullata is like the housefly, though larger. In nature, it feeds upon plant sugars and carrion; in the laboratory, it can be raised on liver. The females usually lay their eggs in the carcasses of dead animals. The larvae (maggots) feed on this material, grow in size through a series of molts, and eventually pupate, forming a cocoon much like that of a moth, except that it consists of the larval epidermis and not of woven material. At room temperature, the egg hatches into a larva in about two days; the larva feeds for eight days and then pupates; pupation lasts for about four days, after which an adult fly hatches out of the pupa. Adults live about two months.

You will be provided with a screen-top glass jar containing enough flies for your experiments. Follow the directions below very carefully to prepare the animals for use in your experiments.

Blowfly preparation

1. Push the jar with the flies into a container of ice until all but the top is covered. Also cool a culture dish to receive the flies after they have been immobilized by the cold. Leave these containers in the ice for ten minutes or more, during which time the poikilothermic flies will cool, become slowly immobilized, and drop to the bottom of the jar. While they cool, go on to step 2.

2. Put some crushed ice in a dissecting pan, put the cooled culture dish on the ice, and place the dissecting pan in a convenient spot for the following manipulations. Before proceeding, wipe away any condensation from the culture dish in the ice container.

3. When the flies are immobilized, place two of them in the culture dish (the bottom of which should be dry) and promptly replace the screened container in the ice. Using a camel's-hair brush, transfer the flies from the culture dish to pipette tips. The flies are mounted in the pipette tips with the head and prothoracic tarsi exposed, as shown in Figure 15-1. The large end

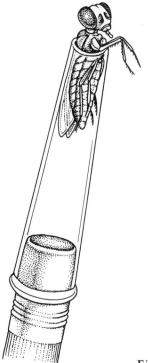

Fig. 15-1. Blowfly mounted on micropipette.

of the pipette can be mounted on a dowel stick or pencil for easier manipulation. Place the prepared fly into a holder.

Repeat this procedure, using a few more flies each time until all have been placed in the pipette tips. (Be very careful to release only a few flies at a time from the stock container, especially when you are handling them for the first time. They warm up quickly and will escape.)

4. Label each stick with a small numbered piece of masking tape, and keep a record of the treatment each fly has undergone. Careful records of this kind will eliminate any question of whether a fly is suitable, by virtue of its previous treatment, for a particular experimental series.

OLFACTORY REACTIONS

Although we shall distinguish **olfaction** (smell) from **gustation** (taste), you should recognize that they are only variations of the same type of sense. Both depend upon contact between chemicals and chemoreceptors, and the distinction between them is based only upon the medium in which the stimulus is carried—gaseous medium in smell; fluid medium in taste. In some situations it is difficult to make this distinction.

Figure 15-2 shows the front view of the head of a housefly, much like *Sarcophaga.* A combination of mouth parts forms the **proboscis**, a strawlike organ used in sucking up juices that the fly feeds on. The proboscis is extended in Figure 15-2. The flattened, aboral end of the proboscis is called the **labellum.** In the following experiments, a positive response to some chemical is characterized by extension of the proboscis to position 2 or beyond (Figure 15-3), while a negative response is characterized by failure to extend the proboscis.

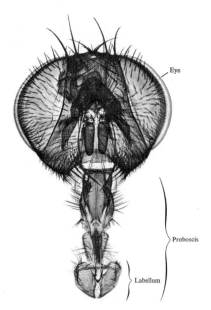

Fig. 15-2. Head of a housefly showing its mouth parts. *(Courtesy Carolina Biological Supply Company.)*

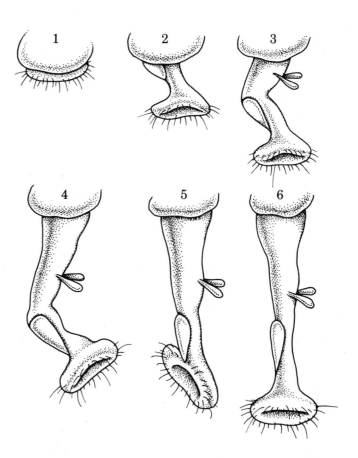

Fig. 15-3. Degrees of extension of the proboscis. *(From V. G. Dethier,* The Hungry Fly, *Harvard University Press, 1976.)*

Testing smell   1. Obtain a rack of labeled test tubes containing a variety of substances with which flies often come into contact. For example, use water, concentrated sugar solution, fresh meat juice, and spoiled meat juice.

Using *separate* wooden sticks for each substance, dip one end of the stick into the solution, and hold it close to the head of each of four flies, being sure that it is safely out of reach of the proboscis and all legs. Record positive and negative results (+ or –) in Table 1, below.

2. Put some drops of water on a clean glass slide or in a petri dish. Allow the forelegs of each fly to touch the water, and let each fly drink its fill (Figure 15-4). Repeat the above tests for olfactory responses with each fly, and record the results in Table 2. Does water satiation have any effect on olfactory response? If so, how might the differences in result in Tables 1 and 2 be explained?

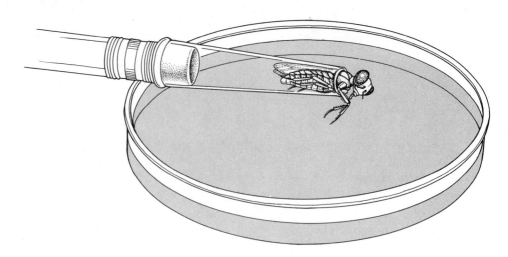

Fig. 15-4. Feeding or watering a blowfly.

FLY NUMBER / TEST SUBSTANCE	1	2	3	4
A				
B				
C				
D				
E				
F				

TABLE 1. OLFACTORY REACTIONS BEFORE DRINKING.

FLY NUMBER / TEST SUBSTANCE	1	2	3	4
A				
B				
C				
D				
E				
F				

TABLE 2. OLFACTORY REACTIONS AFTER DRINKING.

GUSTATORY RESPONSES

Testing taste

1. With four fresh flies not previously used for experiments, test for a taste response to each of the substances used for testing olfactory reactions. Allow the forelegs to contact drops of each substance placed on a slide or in a petri dish. Make sure that you present materials from *behind* the fly (Figure 15-5) so that no part of the head comes into contact with the test material. *Do not* let the flies feed at this point; simply watch for a reaction, and remove the fly promptly after it has occurred or, in the case of no response, after a few seconds. Record the results in Table 3.

2. Repeat the tests for gustatory responses after allowing each of the flies to drink its fill of distilled water. Before beginning the experiments, rinse the forelegs of each fly by passing them through several drops of distilled water. Do this again between each test. Record the results in Table 4.

3. Compare the olfactory and gustatory responses under conditions of thirst and water satiation, and suggest hypotheses to account for the differences observed. Can you be certain, in the gustatory tests, that olfactory responses are not contributing to the results? Is it possible to separate these responses completely?

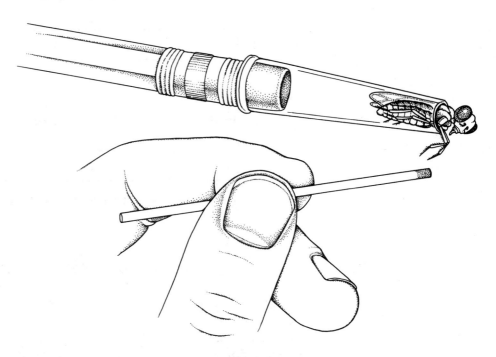

Fig. 15-5. Testing olfactory responses. Be sure to present the potential food materials from *behind* the fly.

FLY NUMBER TEST SUBSTANCE	5	6	7	8
A				
B				
C				
D				
E				
F				

TABLE 3.   GUSTATORY REACTIONS BEFORE DRINKING.

FLY NUMBER TEST SUBSTANCE	5	6	7	8
A				
B				
C				
D				
E				
F				

TABLE 4.   GUSTATORY REACTIONS AFTER DRINKING.

THRESHOLD OF RESPONSE TO SUGAR

As you have probably guessed, taste receptors in adult flies are located on the legs and labellum. When a sugar chemoreceptor in a fly is stimulated, a specific response occurs: the proboscis is extended, and the sugary liquid is imbibed. Some sugars are more effective than others in stimulating the receptors. The stimulatory effectiveness of a compound may in part depend upon its molecular shape. This principle may extend to human chemoreceptors, although there is now conflicting evidence on this point. Some substances, such as salts and alcohols, stimulate different neurons in the sensory hairs and will cause rejection of a previously acceptable sugar solution.

The lowest stimulus effective in producing a response is called the **threshold stimulus**. Thresholds of chemoreceptors may be altered by many factors, such as age, sex, temperature, humidity, and the nutritional state of the fly. In the following experiments, we will determine the threshold for several different sugars, the rejection threshold for salt, and the effect of the nutritional state of the animal on the sugar threshold. If you have time, you may wish to design and carry out other experiments, but first obtain your instructor's permission.

Determining sugar thresholds

The following dilutions of sugars will be available:

	0.005	0.01	0.05	0.1	0.2	0.5	1.0	
Galactose	0.005	0.01	0.05	0.1	0.2	0.5	1.0	MOLAR
Glucose	"	"	"	"	"	"	"	
Sucrose	"	"	"	"	"	"	"	

1. By dipping a clean finger into each solution (from least concentrated to most concentrated) and tasting until you first detect the sugar, find your own threshold to each of the sugars. Enter the results in Table 5.

2. Use depression slides or other small containers to hold the sugars for testing the flies, making sure that each container is properly labeled. With so many concentrations it is very easy to get them confused. Allow each of the experimental flies (which should have been starved for forty-eight hours) to drink distilled water for five minutes, rest for five minutes, again drink for five minutes, and again rest for five minutes. Then test it by bringing the forelegs into contact with each of the solutions, beginning with the least concentrated solution. Record as each fly's threshold that concentration at which the fly first gives a positive response (i.e., proboscis extension). *Do not* allow the flies to feed, and make sure that they give several negative responses to water before you begin tests with them. (Why are these precautions necessary?)

Enter all the results in Table 6. How much variability is there among individual flies? Plot your results on Graph A, using the horizontal axis to

GALACTOSE:	
GLUCOSE:	
SUCROSE:	

TABLE 5. HUMAN SUGAR THRESHOLDS.

FLY NUMBER / TEST SUGAR									
GALACTOSE									
GLUCOSE									
SUCROSE									

TABLE 6. FLY SUGAR THRESHOLDS (BEFORE FEEDING).

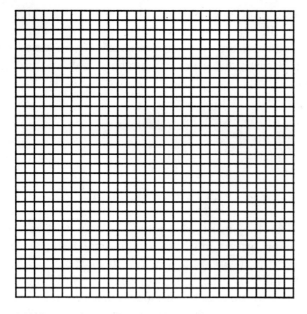

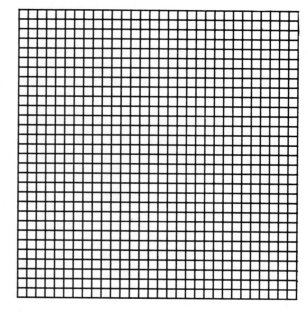

GRAPH A. FLY SUGAR THRESHOLDS (INDIVID-
UAL DATA).

GRAPH B. FLY SUGAR THRESHOLDS (CLASS
DATA).

express sugar concentration and the vertical axis to record the number of
flies that responded to each concentration. Graph A should carry three plots,
one for each sugar. Graph B can be used to plot the results polled from the
entire class.

Would you say that your flies represent an adequate statistical sample?
Do the class data form a better sample? Why?

Compare sensitivity of human and fly to the different sugars, considering the threshold to each sugar as the lowest concentration to which at least
50% of the flies (or humans) respond.

Determining repellents   The threshold concentrations of substances that flies find repellent (such as
salt, NaCl) may be determined by combining such substances with a solution
known to be acceptable and above threshold. Failure to extend the proboscis
to a solution in which sucrose, for example, is above threshold is an indication that the repellent is present in above-threshold amount.

Select one of the above-threshold concentrations of sucrose to which a
fresh fly consistently responds with acceptance. Do not, however, allow the
fly to feed—remove the fly at once after the positive responses have taken
place.

Mix an equal number of drops of the selected sucrose solution with
drops of the repellent solution (1.0 M NaCl or some other substance supplied
by your instructor). Test the fly's response to this mixture, and record the
results in Table 7. Repeat the experiment, using a lower concentration of
NaCl* if the fly rejected the mixture, or a higher concentration of repellent

*The concentration may be halved, for example, by mixing a drop of 1.0 M NaCl with a drop
of water.

FLY NUMBER REPELLENT SOLUTION										

TABLE 7.   REJECTION THRESHOLDS.

(e.g., 2.0 M NaCl) if the fly did not reject the first mixture. Try the first repellent solution with higher concentrations of sucrose. Does the rejection threshold depend entirely upon the concentration of repellent?

Place a drop of sucrose solution very close to a drop of salt solution. Position a fly so that its tarsi (legs) touch one solution and its proboscis is over the other solution. Can you fool the fly into feeding on the repellent substance? Does the perception of salt by the tarsal receptors inhibit feeding on the sugar? Design an experiment to study the effects that stimulation of one set of receptors has upon another. How would you determine if the absence of tarsi has any effect upon feeding habits?

Effect of nutrition on the sucrose threshold

Select several flies, and allow each to feed on the 1.0 M sucrose for ten to fifteen seconds. In the usual manner, determine the threshold of each fly to each of the test sugars. Record the data in Table 8, and compare it with your original data for threshold in starved flies. Explain the results, and suggest hypotheses for the differences observed.

FLY NUMBER REPELLENT SOLUTION										

TABLE 8.   FLY SUGAR THRESHOLDS (AFTER FEEDING).

THE EFFECTS OF FOOD ON LOCOMOTION

You will be supplied with several flies from which the wings have been removed, or your instructor may ask you to remove the wings yourself with fine scissors. These flies have been fed 0.1 M sucrose, a minimal diet, and have been allowed to get used to their condition overnight.

Locomotion behavior

1. Place one of the flies on the table, and watch its pattern of walking. Does it walk in straight lines, or does it deviate from straight lines?

2. Place a drop of above-threshold (1.0 M) sucrose in contact with the fly, or bring the fly into contact with the drop. Remove the animal immediately from the drop, and compare its walking pattern with the walking pattern of the first test. If no difference is noticed, return the fly to the stock container and try the experiment with another fly.

Can you suggest an explanation for this difference in behavior? How might this be an advantage to the animal in nature?

3. With a fine brush, draw thin, concentric circles of distilled water and sugar solution on white paper, as shown in Figure 15-6. Draw one pattern at a time.

> a. Place a wingless fly in the innermost circle of the first set (A), and watch its behavior continuously until it walks out beyond the outermost circle. Record the walking with a fine line penciled into Figure 15-6.
>
> b. Repeat the procedure in step 1 with circles B and C (Figure 15-6), using a fresh fly each time. Record the walking, as before.

Explain the behavior in each case.

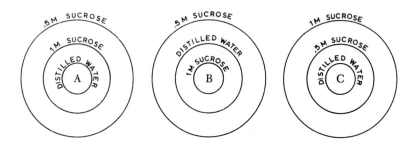

Fig. 15-6. Circles used for studying the effects of food on locomotory behavior.

4. Using 1.0 M sucrose, draw a "Y" about ½ inch long on a piece of white paper. Place a fresh fly at the stem of the "Y," and let it walk. Carefully record its movements, and watch especially what happens when the fly reaches the branching point of the "Y." What determines the direction the fly chooses at the branching point? Repeat the experiment several times until you have the answer.

5. On a piece of white paper, draw an "M" in 1.0 M sucrose. Put a wingless fly at one end of the "M," and let it walk. Can the fly follow the trail? If so, have you trained it, or has it "learned" to behave in this fashion?

<table>
<tr><td></td></tr>
</table>
Biological Science, 4th ed.
Chap. 23, pp. 615–643
Elements of Biological Science, 3rd ed.
Chap. 21, pp. 391–407

TOPIC 16 CELLULAR REPRODUCTION

Though individual cells and organisms die, the continuity and increase of life are guaranteed by two types of nuclear division: **mitosis** and **meiosis,** processes that differ principally in the behavior of the chromosomes.

Coupled with cytoplasmic division processes, mitosis is a mechanism for transmitting unaltered genetic constituents from one generation of cells to another. This is ordinarily accomplished through a duplication of chromosomes, so that two complete sets are present in the parent cell; after the duplication, the sets are separated in such a manner that each daughter cell inherits a complete chromosomal complement, identical to that of the parent (Figure 16-1A). Mitotic cell divisions are important in growing and regenerating regions of organisms; in plants, they are confined mainly to the

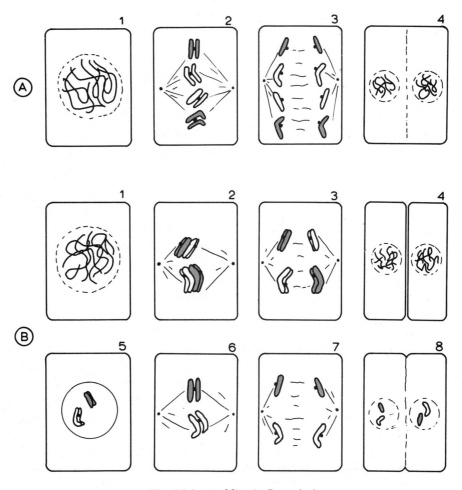

Fig. 16-1. A: Mitosis; B: meiosis.

meristems, such as buds, root tips, and cambium. In animals, mitotic divisions are found in cells at the base of epithelia, in the bone marrow, and in many other regions where tissues are growing or cells are being replaced.

Meiosis, in contrast, is a mechanism for reducing the number of chromosomes by half. In animals, this takes place in cells that will give rise to gametes, but it may occur at other points in a sexual reproductive cycle, as it usually does in plants. To understand meiosis, recall that chromosomes commonly occur in pairs; members of a pair carry genetic factors for the same traits and are called **homologous chromosomes**. During *mitosis,* homologous chromosomes behave *independently,* and a copy of each is transmitted to each daughter cell. During *meiosis,* however, the homologues associate or **synapse**; synapsis is followed by separation of the members of each homologous pair into *different* daughter cells (Figure 16-1B). This fundamental difference in chromosomal behavior—independence of homologous chromosomes in mitosis and association and separation of homologous chromosomes in meiosis—is the key to understanding the differences between these two otherwise mechanically similar processes.

In this unit, you will make and study a slide of dividing cells in the root-tip meristem of broad bean, *Vicia faba.* You will have an opportunity to review your textbook understanding of cell division and a chance to observe firsthand the behavior of chromosomes in mitosis. A clear understanding of chromosomal behavior in both mitosis and meiosis is essential to your later work in genetics.

TECHNIQUE FOR MAKING SLIDES

Prior to the laboratory period, seeds of the broad bean were germinated and the root tips, which contain many actively dividing cells, were cut off, killed, and hardened in an acid solution. Preparation of this material for microscopy consists of three steps:

1. The root tip is prepared for staining by treating it with hydrochloric acid, which exposes chemical groups called **aldehydes**, present in the chromosomal DNA, to which the stain may then attach.

2. The root tip is stained with **Feulgen reagent**, a dye mixture that adheres to the aldehyde groups exposed in step 1, thereby coloring the chromosomes a deep magenta.

3. The cells are dispersed into a single layer on a slide, and the slide is prepared for microscopic observation. The slide may be studied at once, or it may be run through several other steps to make it permanent.

Making slides is not a haphazard process, but requires a great deal of care if the resultant preparation is to be worthwhile for study. Since you will have only one opportunity to make a good slide, read the directions carefully before you begin and follow them to the letter, noting any modifications your instructor may suggest at the beginning of the period.

Directions for temporary mounts

1. From your instructor, obtain a root tip in a vial of fixative. Discard the fixative into the waste beaker at your desk. Add distilled water to the

vial, and leave the vial undisturbed for one minute. Discard the water, and add 6 N hydrochloric acid to half the capacity of the vial. Note the time. Leave the acid on the roots for three minutes, and then, with a dropper, remove the acid and discard it into the waste beaker. Add distilled water to the vial to rinse the roots. Agitate the vial gently for one minute, and discard the water.

2. With forceps, place the roots in a second vial, containing Feulgen reagent. Note the time, and remove the roots after twenty minutes. While the roots are staining, clean the first vial, and partially fill it with 45% acetic acid. Put the roots in the acid when they are brought out of the stain.

(*Note:* If you wish to make a permanent slide, omit step 3 and follow the steps in the following section. Read all the steps in that procedure before you try it.)

3. With insect pins or a razor blade, tease or chop the tissue into very small bits in a drop of 45% acetic acid on a clean slide. This operation damages a few cells but makes the majority much easier to smear later on.

Adding a drop of acetic acid if necessary, gently put a cover slip in place, taking care to eliminate air bubbles. Put a small clean cork upside down on the cover slip, and press down firmly in a vertical direction (Figure 16-2). Properly performed, this operation will smear the cells into a very thin layer. If the pressure is vertical, it is almost impossible to break the cover slip; it takes a great deal of pressure to make a proper squash, so don't hesitate to apply it.

Fig. 16-2. Technique for making a tissue squash. Tease the tissue into small pieces; add a drop or two of liquid, a cover slip, and a cork; and apply vertical pressure.

Additional directions for permanent mounts

1. Clean a slide and cover slip with 95% alcohol, and place both of these items on a clean paper towel at your desk. Put one drop of Mayer's albumen on the slide. With a clean finger, wipe almost all of it off so that only a thin film remains. It is *essential* that a *clean, thin* film cover the middle third of the slide. If this doesn't work the first time, clean the slide and try again.

2. Follow step 3 in "Directions for Temporary Mounts," above, for making the smear.

3. Place the slide on a piece of dry ice for five to ten minutes. At the end of this time, while the slide is still on the dry ice, insert a razor blade at the edge of the cover slip and flip the cover slip off. Immerse the slide *at once* in a Coplin jar of 95% alcohol, where it should remain for five minutes.

4. Run the slide through three changes of 100% alcohol, each lasting five minutes, and one change of xylene, also lasting five minutes.

5. Remove the slide from the last solvent change, let the excess solvent run off the slide onto a towel (but don't let the slide dry), and promptly place a drop of Euparal, Damar, or other mounting medium over the cells. Using your best technique to avoid air bubbles, mount the clean cover slip over the Euparal or Damar. If there are bubbles in the medium before the cover slip is applied, they may be removed by poking them with a heated needle.

IDENTIFICATION OF MITOTIC PHASES

Although cell division is commonly separated into "phases" for convenience of discussion, you should remember that such separation is arbitrary and that the process is in fact continuous. Keep in mind as you examine your slide that the stain you have used colors only DNA; other structures will show up indistinctly, if at all. The directions below refer to the broad bean, in which you will see only the chromosomal events in cell division; other events, especially those involved in cytoplasmic division (**cytokinesis**) are much more conveniently studied in prepared slides, described in the section using onion and whitefish.

Using a compound microscope, study your slide under both low power and high power, and refer to Figure 16-3.

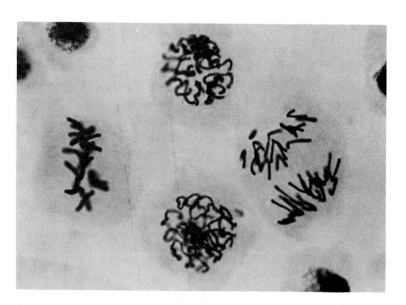

Fig. 16-3. Mitosis in hyacinth. This photograph was prepared from a squash mount similar to that you will make from *Vicia faba* root tip. *(Courtesy Carolina Biological Supply Company.)*

Broad bean

1. A majority of the cells on your slide will be in **interphase**. The stained granular nucleus of such cells is clearly delineated in the cytoplasm, though individual chromosomes are not visible. Often two or three nucleoli will be

visible. Are they stained? What events take place during interphase? What evidence is there that they occur?

2. During **prophase**, the chromosomes—now doubled so that each consists of a pair of identical **chromatids**—shorten and thicken. Although they were not visible as discrete bodies during interphase, the chromosomes now appear as tangled threads. Later, as the threads become more distinct, the nuclear membrane disappears. Compare the appearance of the chromosomes in early and late stages of prophase. Note that in the former stages, the chromosomes are clearly enclosed in a spherical area—the nucleus. Later, after the nuclear membrane has disappeared, the chromosomes spread out.

The pair of chromatids composing each prophase chromosome is attached at one point, the **centromere**. The centromere, in turn, apparently is the point of attachment of the chromosome to the **spindle**. The spindle (not visible on your slide but easily seen in prepared slides), formed in early prophase, consists of protein fibrils that probably play a key role in moving the chromosomes during the division.

By the end of prophase, the chromosomes are very short and thick, and have migrated to the center of the cell.

3. During the brief stage termed **metaphase**, the chromosomes are arranged along the equatorial plate* of the spindle and appear to form a line across the middle of the cell. Each chromosome is attached by its centromere to a fibril of the spindle. (Under high power, the centromere is visible in some metaphase chromosomes as a clear area in the chromosomes.) It is actually the centromeres that are lined up precisely along the equatorial plate, while the "arms" of the chromosomes may extend toward the poles in various directions, giving the chromosomes a "frayed" appearance. At the end of metaphase, each centromere divides; each of the former chromatids now has its *own centromere* and is designated a separate chromosome.

4. At the beginning of **anaphase**, the two new single-stranded chromosomes move away from one another, one going toward one pole of the spindle and the other to the opposite pole. You can recognize a cell in early anaphase by observing that the chromosomes are in two groups, a short distance apart. By late anaphase, each chromosome group has almost reached its respective pole.

Under high power or with an oil-immersion objective, compare the appearance of metaphase and anaphase chromosomes. If possible, locate these two phases close together on your slide. Make sure that your lenses are very clean and that the light is properly adjusted. Under favorable conditions, you will see that a metaphase chromosome consists of two chromatids connected by the unstained centromere, while an anaphase chromosome is single-stranded. If you cannot see the double structure of the metaphase chromosome, a comparison of the width of an anaphase and a metaphase chromosome in adjacent cells should demonstrate clearly that metaphase and anaphase chromosomes consist of different amounts of material.

5. In early **telophase**, the chromosome clusters have reached the poles, but the "arms" of the chromosomes are still oriented along the spindle. In mid-telophase, the clusters are more compact, and individual chromosomes are no longer oriented along the spindle. By late telophase, new nuclear

*An imaginary plane running through the middle of the cell at right angles to the long axis of the spindle.

membranes have formed around the clusters, and individual chromosomes are indistinct. Telophase ends when the nucleus assumes the characteristics of interphase, and the full mitotic cycle is completed.

When you have gained a clear idea of chromosomal behavior in mitosis, find examples on your slide that *differ* from the illustrations, and make a large drawing of the nucleus and/or chromosomes of each phase. It may be necessary to make several different sketches for some of the phases that change in appearance as they progress. Label the drawings carefully.

Onion and whitefish (Figures 16-4 and 16-5)

1. Locate the five stages of mitosis in a prepared slide of onion (*Allium*) root tip (longitudinal section). Nucleoli will be clear in interphase cells, and the spindle is usually seen easily in anaphase cells. Cytokinesis begins during early telophase and is generally complete before telophase has finished. As in most plants, cytokinesis occurs through the formation of a **cell plate** (which eventually becomes the middle lamella of a new cell wall) in the region of the equatorial plate.

2. Locate the five stages of mitosis in a prepared slide of whitefish (*Leucichthys*) blastula. A blastula is an early embryonic stage produced by successive early divisions of the zygote. *A word of caution:* When you study these slides, remember that the cells of the blastula are not oriented in the regular longitudinal fashion typical of a plant root. Therefore, a section through the whole embryo cuts cells in many different ways. Try to locate longitudinally sectioned cells to study mitosis.

Animal cells contain a **centriole**, a structure associated with spindle formation that is generally absent in plant cells. The centriole is too small to be seen without the electron microscope, but it is located in the centrosome, a clear area that can be easily seen. Some of the fibrils that radiate from the centriolar region are fibrils of the spindle; others, radiating toward the poles, form two starlike **asters**, also absent in plant cells.

Cytokinesis occurs as a constriction, or **cleavage furrow**, which forms in the equatorial region. This furrow deepens progressively until it cuts through the cell and its spindle, producing two new cells.

Make sketches to show the differences between the plant and animal cells in mitosis and cytokinesis, or label Figures 16-4 and 16-5.

THE TIME SEQUENCE IN MITOSIS

In your slide examination, you probably noticed that some stages were found more often than others. Which phase did you see most frequently? There are differences in the frequency of appearance of the phases because each phase takes a different amount of time to occur. The time a cell spends in interphase, plus the time spent in mitosis, is referred to as the cell cycle.

Considering only what you know about the events of each phase and *without* consulting your slide, make a list of the five phases in order of *decreasing* frequency in which you would expect to find them.

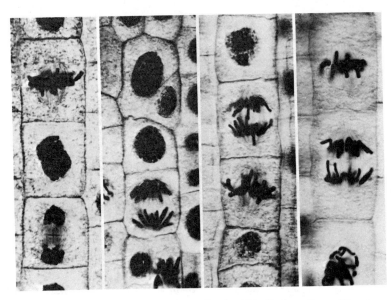

Fig. 16-4. Mitosis in onion root tip—longitudinal section. *(Courtesy Carolina Biological Supply Company.)*

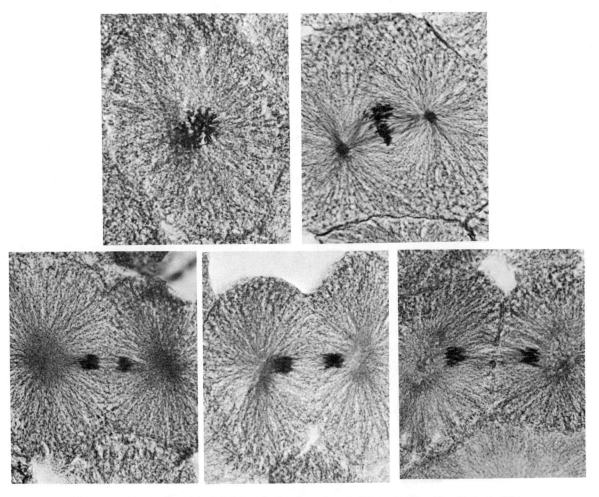

Fig. 16-5. Mitosis in whitefish blastula (early embryo). *(Courtesy Carolina Biological Supply Company.)*

1.
2.
3.
4.
5.

Now examine the slide, and, in the following manner, determine the relative frequency of each of the five stages in mitosis:

1. Select a region of the root tip that seems to have dividing cells scattered regularly throughout it and is one cell thick.

2. In this field, count the number of cells in each phase, entering that number in the chart below and in the chart for data from the entire class. Add the numbers in your column to find the total number of cells counted.

3. Divide the number of cells in each stage by the total number counted, and multiply by 100 to get the percentage of cells in each stage.

	NUMBER OF CELLS	PERCENTAGE OF CELLS	TIME
Interphase			
Prophase			
Metaphase			
Anaphase			
Telophase			
Your total			
Class total			

4. Assume that the differenece in amount of time for completion of each stage accounts for the difference in observed frequency of stages. Since mitosis in a broad bean normally takes about sixteen hours from the beginning of interphase until the end of telophase, calculate from your percentage data the approximate time (in minutes) it takes for the completion of each phase of mitosis. Enter these calculations in the appropriate column of your table.

Which stage is the longest? Is this what you predicted?

Which stage is the shortest? Does this comply with what you know occurs during this stage?

Compare the frequencies of stages in the class data with your own. If there are discrepancies, how could you explain them?

Why do you think it is better to draw conclusions from the sum of data submitted by the class than from your data alone?

QUESTIONS FOR FURTHER THOUGHT

1. In a single sentence, how would you explain the basic significance of mitosis?

2. Explain the difference between the terms "chromosome" and "chromatid."

3. In which stage of mitosis does DNA duplication occur?

4. Suggest a hypothesis to explain the adaptive value of having genetic material packaged in chromosomes.

5. What evidence is there in anaphase cells to indicate that chromatids are mechanically moved apart during anaphase? What structures are probably responsible for the movement?

6. Which of the phases do you think would require the greatest expenditure of energy? How would you design an experiment to find out?

7. If you know how long it takes to complete a mitotic cycle, and if your information regarding the time for completion of each phase is correct, what other information would you need in order to calculate the *rate of movement* of chromosomes in anaphase?

8. You inferred the length of mitotic phases from information about the total time for a mitotic cycle and the relative frequency of each stage. What technique would you need to get the same information directly?

9. In a single sentence, state the fundamental difference in chromosomal behavior between mitosis and meiosis.

10. Without looking at your notes or your laboratory manual, take a sheet of paper and make large diagrams of mitosis and meiosis in an organism with three pairs of chromosomes. Make a list of the most important differences between mitosis and meiosis.

11. Define the following terms, or, where appropriate, describe their role in cell division:

prophase	cell plate
metaphase	nucleolus
anaphase	centromere
telophase	aster
interphase	homologous chromosomes
cytokinesis	synapsis
cell cycle	

17 GENETICS AND THE ANALYSIS OF DATA

Biological Science, 4th ed.
Chaps. 24, 25
Elements of Biological Science, 3rd ed.
Chap. 14

The chromosome theory of heredity is based largely upon experiments that involved the crossing, or mating, of diploid organisms, those that contain pairs of alleles for each trait. In heterozygotes, the presence of one kind of allele often totally masks or dominates the other; therefore, the actual genetic makeup of the organisms can be ascertained only through breeding experiments in which the alleles are *segregated* into gametes and then *recombined* in different ways. In this laboratory, you will study the results of some crosses that are similar to those performed by Mendel in his classical experiments on peas. In part, this study will enable you to determine what kind of genetic information can be inferred from such experiments. In addition, you will learn how the data resulting from such experiments can be subjected to statistical tests of validity. Finally, you will perform some crosses yourself and will examine a genetically controlled trait in humans.

MONOHYBRID AND DIHYBRID CROSSES

Monohybrid Cross A monohybrid cross seeks to determine the pattern of inheritance of a pair of contrasting alleles. The production of chlorophyll in tobacco, for instance, is controlled by a dominant allele, which we may arbitrarily designate *C;* the alternative recessive allele *c,* when present on both chromosomes, gives rise to albino plants. In Experiment A, you will be provided with a dish of agar upon which about 100 tobacco seeds have been germinated. These seeds were parented by a pair of heterozygous (*C/c*) tobacco plants. (Alternatively, your instructor may ask you to plant the seeds yourself one week prior to the laboratory.)

Classifying and counting tobacco seedlings

Classify and count the seedlings, and record the numbers in Table 1, below. Had you not been given the information about dominance of the allele for chlorophyll production, how would you have been able to determine that information from your data?

| | NUMBER OF SEEDLINGS | |
	GREEN	ALBINO
Individual		
Class		

TABLE 1.

What crosses would you have made to produce the hybrid parents of the seeds you germinated? What would have been the genotype of each of the parent plants?

What is the genotype of albino plants?

What are the possible genotypes of your green plants? If you had to obtain more hybrid plants in order to get seeds for next year's class, how would you isolate the hybrids from your own plants?

Statistical analysis of data

Based on your knowledge of segregation and recombination of alleles, you would have predicted a *phenotypic* ratio of three greens to one albino as shown in the Punnett square below:

	C	*c*
C	*C/C*	*C/c*
c	*C/c*	*c/c*

The ratio you actually found when you counted the seedlings was probably not exactly 3:1, just as flipping a coin a number of times will not yield exactly equal numbers of heads and tails, even though the predicted ratio is 1:1. If the *deviation* (*d*)—the difference between your actual and expected (*e*) results—is small, you would not be very disturbed; intuitively and correctly, you expect some deviation in any random process, such as coin tossing or, in the tobacco experiment, the segregation of alleles at meiosis and their recombination at fertilization. But what if the deviation were large? Was something wrong with the way the experiment was run? Was something wrong with your hypothesis (what you expected)? How large must a deviation become before you have to reject your hypothesis and form a new one?

The problem of formulating mathematically consistent and objective answers to such questions faces almost every scientist and enters almost every realm of research. The use of statistics offers a reasonable and standardized solution. One statistical test of deviation significance, widely applied to genetic ratios, is the **chi-square** (χ^2) **test**. The formula for this test, derived from reliable mathematical calculations, is:

$$\chi^2 = \Sigma (d^2/e),$$

where *d* is the deviation from the predicted value, *e* is the expected (hypothetical) value, and Σ means "the sum of." The expression (d^2/e) alone gives a chi-square value for each "class" or group of data in which deviation can occur. For example, in our tobacco experiment there are two classes of data, green and albino. As we shall see, chi-square values for each group of data are added, and the total value is compared with a table that gives information about the significance of the deviation.

To illustrate the use of this test, consider two hypothetical tobacco-seedling experiments. Suppose that in Experiment A, we had obtained 85 green seedlings and 35 albinos. The total sample would then have been 120, and the expected ratio would have been 3:1 or 90 greens to 30 albinos. The data and chi-square calculations for Experiment A are in Table 2. In hypothetical Experiment B, suppose we had counted a total of 32 seedlings, 19 of which were green and 13 albino. Fill in the corresponding information and calculation for Experiment B in Table 3.

	CLASS 1	CLASS 2
Observed values	85	35
Expected values (e)	90	30
Deviation (d)	−5	+5
Deviation squared (d^2)	25	25
d^2/e	$25/90$	$25/30$
$\chi^2 = \Sigma(d^2/e) =$		$0.28 + 0.83 = 1.11$

TABLE 2. EXPERIMENT A.

	CLASS 1	CLASS 2
Observed values		
Expected values (e)		
Deviation (d)		
Deviation squared (d^2)		
d^2/e		
$\chi^2 = \Sigma(d^2/e) =$		

TABLE 3. EXPERIMENT B.

Notice the difference in total chi-square value: about 1.1 for Experiment A and 4.1 for Experiment B. Let us consider three important factors that influence the size of these numbers.

First, from the formula for the chi-square test, it should be clear that the larger the value of d, the larger the chi-square value for that class of data. Second, the larger the value of e (expected), the smaller the chi-square value for that class of data. Finally, the more classes of data under consideration, the larger the total value. For example, if we had only two classes of data (as in our tobacco experiments), we expect less opportunity for deviation (and a smaller chi-square total) than if we had considered an experiment in which there are ten different data classes and, thus, ten opportunities for deviation. Since the d value and the number of classes of data were the same in both of our hypothetical experiments, only the expected values varied. These reflect the size of the sample. In a sample as small as that in Experiment B, we should not be surprised to find a large chi-square value.

Chi-square is, therefore, a reflection of three factors: deviation size, sample size, and the number of opportunities in which deviation can occur. It can help us to answer two questions:

1. How probable is it that our chi-square values would have occurred simply by chance?

2. Should we reject our hypothesis or not?

We can answer question 1 by comparing our chi-square values with those shown in Table 4.

Table 4 shows the maximum chi-square values allowable if the deviations actually found are to be considered due to chance alone. The vertical column shows the number of classes of data under consideration minus one, a number called the degrees of freedom. For both of our hypothetical Experiments A and B, this number is 1. The *P* value, on the other hand, expresses the *probability* that deviatons yielding the indicated value are due to chance alone. Comparing our value of 1.11 (for Experiment A) with Table 4, we find that the corresponding *P* value for one degree of freedom is between 0.20 and 0.30. This means that we could expect a chance deviation as large or larger than the one we actually obtained in 20% to 30% of the cases considered, if we were to repeat the experiment again and again.

	NO REASON TO DOUBT HYPOTHESIS							REASON TO DOUBT HYPOTHESIS		
	DEVIATIONS INSIGNIFICANT							DEVIATIONS SIGNIFICANT		
P / Degrees of freedom	.99	.95	.80	.50	.30	.20	.10	.05	.02	.01
1	.00016	.0039	.064	.455	1.074	1.642	2.706	3.841	5.412	6.635
2	.0201	.103	.446	1.386	2.408	3.219	4.605	5.991	7.824	9.210
3	.115	.352	1.005	2.366	3.665	4.642	6.251	7.815	9.837	11.341
4	.297	.711	1.649	3.357	4.878	5.989	7.779	9.488	11.668	13.277
5	.554	1.145	2.343	4.351	6.064	7.289	9.236	11.070	13.388	15.086

TABLE 4. VALUES OF CHI-SQUARE.

To answer question 2, we must first decide on a reasonable *P* value and then find where our chi-square total lies with respect to that *P* value. A high chi-square value usually indicates that the experimental results deviate greatly from the predicted results. The more the results deviate from the predictions, the more likely it is that the predictions were based on an incorrect hypothesis. It follows, therefore, that a high chi-square value is a more significant indicator of the possible need for revision of the hypothesis than is a low chi-square value. By convention, we accept as statistically significant (and, therefore, as cause to doubt our hypothesis) any deviation so great that the probability of its having occurred by chance alone is 0.05 or less. In other words, any chi-square value large enough to yield a *P* value as low as 0.05 is said to reflect a statistically significant deviation—a deviation that warrants re-examination of the hypothesis.

A chi-square value large enough to yield a *P* value of 0.01 is said to be highly significant.

On this basis, the deviation in Experiment A was insignificant, and the deviation in Experiment B significant. Had we actually obtained experimental values like those in hypothetical Experiment B, we might have repeated the experiment with a larger sample, recognizing that the chi-square test is particularly sensitive to sample size. Had we continued to get significant *P* values with a larger sample, we would have had to revise the hypothesis from which our expected values were derived, or we would have had to look carefully at the way we ran our experiment. If we were sure of the hypothesis (as we are in this case), then we may have failed to count the seedlings accurately, or someone may have pulled some of them up before we got a chance to examine them, or perhaps some physiological factor caused differential germination. Only after ruling out such possibilities would we begin to look for a genetic explanation of our significant deviation.

Carry out the chi-square test on your results from the tobacco-seedling experiment, which you recorded in Table 1 (a space is provided in Table 5). Are the deviations statistically significant?

	CLASS 1	CLASS 2
Observed values		
Expected values (e)		
Deviation (d)		
Deviation squared (d^2)		
d^2/e		
$\chi^2 = \Sigma(d^2/e)$		

TABLE 5.

Examining characteristics of the dihybrid cross

Dihybrid Cross   You will be given an ear of corn that represents the F_2 generation of a cross between plants differing in two characteristics. The F_1 offspring were genotypic hybrids for two characters: kernel *color* and kernel *texture*. One pair of alleles controls color (purple or yellow), and the other pair controls texture (smooth or rough, often referred to respectively as "starchy" or "sugary"). The manner in which such crosses are made is shown by the photographs in Figure 17-1.

Classify and count the F_2 progeny (the kernels), and record the data in Table 6.

Which of the alleles for color is dominant? Which of the alleles for texture is dominant? How do you know?

How do you know that the color and texture alleles are located on different chromosomes and are, therefore, segregated and inherited independently?

	CLASS 1	CLASS 2	CLASS 3	CLASS 4
Observed values				
Expected values (e)				
Deviation (d)				
Deviation squared (d^2)				
d^2/e				
$\chi^2 = \Sigma(d^2/e) =$				

TABLE 6.

Fig. 17-1. Controlled pollination of corn. A: Bag is secured over tassel, where pollen is produced; B: bag is placed over ear while it matures; C: bag is removed from mature ear, exposing silk brush; D: pollen is poured from tassel bag over silk brush. The silks are elongated styles (part of the female flower); the pollen lands on the silks and eventually grows down through the silk into the ovary, within each kernel of the ear. *(Courtesy Carolina Biological Supply Company.)*

Determine chi-square values for your own data and then again for data collected from the whole class. Are the deviations significant? If anyone's results show significant deviations, try to determine the cause of these results.

SEX LINKAGE IN *DROSOPHILA*

The fruit fly, *Drosophila melanogaster,* has been a very important experimental organism for genetic studies. Like the other classical subjects of genetic research, it has many easily distinguishable mutants. Its main attractions, however, are small size, short life cycle, and ease of culture in the laboratory, where it can be readily induced to form large numbers of offspring in a relatively short time.

As in many other animals, one pair of chromosomes controls the expression of sex in *Drosophila.* These chromosomes are arbitrarily designated X and Y. A female is genotypically XX, a male XY. Certain traits, such as eye color in *Drosophila,* are carried only on the X chromosome and are spoken of as being sex-linked. Sex linkage was discovered with *Drosophila* in 1910 by Thomas Hunt Morgan, and we shall repeat the classical experiments to determine the phenotypic ratios obtained when alleles (in our case, for eye color) are sex-linked.

The entire experiment spans four weeks, according to the following schedule. Your instructor will explain modifications in the schedule, which parts you are to follow, and when you are to do them.

Week 1. Remove parental generation. Learn how to handle and sex flies, and observe stages in the life cycle (this may also be done in Week 2 or 3).

Week 2. Remove and count F_1 (hybrid) generation. Prepare a second cross using the F_1 hybrids as parents.

Week 3. Remove the F_1 parents.

Week 4. Count F_2 generation. Analyze data.

Week 1 During this laboratory, your instructor will give you your culture so that you can remove the parent flies and learn about the life cycle and handling of the flies. Alternatively, he or she may remove the parents for you and return the culture the following week.

Drosophila *life cycle and natural history*

Drosophila occurs naturally on rotting fruits; both the adults and the larvae feed on plant sugars and on the wild yeasts that grow in rotting fruit. The female flies lay eggs on the same materials. After a day or two, the eggs hatch into small larvae, which feed and grow for about eight days, depending upon the temperature. During the period of growth, they molt twice, resulting in **instars**—three larval periods of growth between molts. When fully grown, third-instar larvae cease feeding, climb onto the walls of the culture vessel, and pupate. During pupation, which lasts about four days, the larval tissues are reorganized to form those of the adult. The life cycle is summarized in Figure 17-2.

All the stages should be visible either in the medium or on the walls of the culture vial. Use a dissecting microscope to find them.

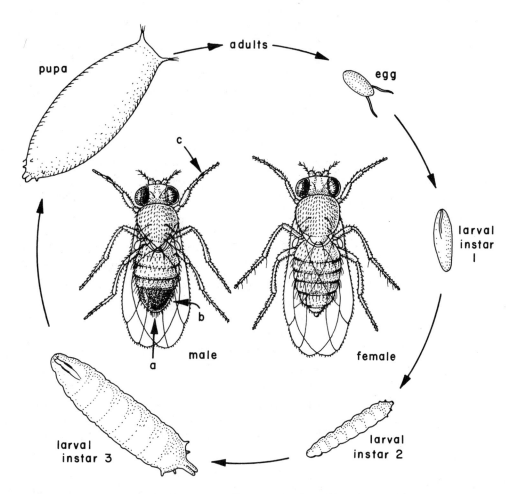

Fig. 17-2. Life cycle of a *Drosophila*. Features distinguishing the adult male are indicated by a, b, and c.

Handling the flies

To facilitate handling and sexing, flies are usually anesthetized with ether. It is important to avoid too much or repeated anesthesia because it easily kills or sterilizes newly emerged flies and has deleterious effects on older flies. The following directions apply to the anesthetizer made by the Carolina Biological Supply Company, but your instructor may prefer to have you use some other method.

1. Remove the cap from the top of the anesthetizer. Remove the cap from the bottom. Fill a third of the hollow stopper with ether for anesthesia. Pour the ether on the foam pad in the bottom of the etherizer. Immediately replace the cap and put the hollow stopper back in the top of the etherizer.

KEEP BURNERS AND CIGARETTES OUT OF THE LAB; ETHER IS HIGHLY FLAMMABLE.

2. When you are ready to etherize the flies, remove the stopper from the top of the etherizer. Tap the bottom of the culture vessel against the palm of your hand to knock the flies down. Remove the plug from the culture. Invert the culture over the etherizer, and tap the flies into the chamber. After all the adults have been tapped into the chamber, quickly right the culture vessel so that its base covers the top of the etherizer. Plug the culture vessel. Tap the base of the etherizer on the table, remove the

culture vessel, and plug the etherizer with its own stopper. Watch the be-havior of the flies in the chamber. About twenty seconds after they stop moving, they may be dumped onto a white card for examination. (*Note:* If it is necessary to reanesthetize flies, the anesthetizer may be reinverted over them with the hollow stopper removed. As noted above, avoid doing this repeatedly.)

Flies are best observed and sorted on a white card, using a camel's hair brush as manipulator.

3. Remove the parent flies from your culture, and purposely kill them by overetherizing. They may then be observed more closely (for sexual characteristics) without fear of their escape.

Sexing Drosophila

It is very important for you to be able to distinguish easily the sex of your flies. There are several sharp anatomical differences that make this a simple matter. In Figure 17-2 some of the sexual differences are indicated by arrows on those parts of the male specimen that differ from corresponding regions on the female specimen.

The male is generally smaller than the female, but this may be confusing if both animals are immature. One good indicator of sex is the presence of black bristles (a) about the male genitalia; these are absent in the female. Another good indication is the dark-tipped male abdomen (b), but this again may be confusing on immature specimens. The sex combs (c), groups of black bristles on the male forelegs, are a good indicator of sex. A final criterion is the distance between the eyes, which is less in the male than in the female; you can make this comparison if the animals are seen together.

You will notice that some of the flies have red eyes and some white eyes. These traits are controlled by a pair of sex-linked alleles. By making reciprocal crosses (red male × white female, and white male × red female), we can learn how these alleles are inherited and what the dominance relation is between them. In Table 7, in the box labeled "Parents," note which culture you received and the number of adults of each sex and eye color that started the culture. This information already will have been recorded on the culture vial.

Discard the dead parent flies—and any other flies you subsequently wish to remove permanently—into a jar of alcohol marked "Morgue" on the supply desk.

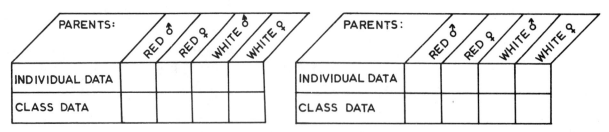

TABLE 7. F₁ *DROSOPHILA*.   TABLE 8. F₂ *DROSOPHILA*.

Salivary glands

In *Drosophila* and some other flies, larval growth occurs principally as a result of cell enlargement. In some species the larval salivary glands consist of large cells containing nuclei with extremely large chromosomes. Such chromosomes have been widely used in studies relating chemical activity in the larva to the morphology (particularly the banding pattern) of the chromosomes.

To see the giant chromosomes, select from your culture a third-instar larva that is crawling on the sides of the culture vessel, preparing to pupate. Place the larva in a drop of saline solution on a microscope slide, and examine the larva with the dissecting microscope. Orient the larva with dissecting needles so that the anterior end, which bears small, black toothlike structures, points away from you. Hold the larva by placing a needle across its posterior end, and insert another needle just behind the mouth parts. Pull the anterior needle away from you, while keeping the posterior one in place. In the resulting debris, you will find the salivary glands—two saclike structures (Figure 17-3). They are almost transparent. Attached to them are small pieces of yellowish, opaque fat tissue. In most cases the glands will have been detached from their common duct.

Salivary glands

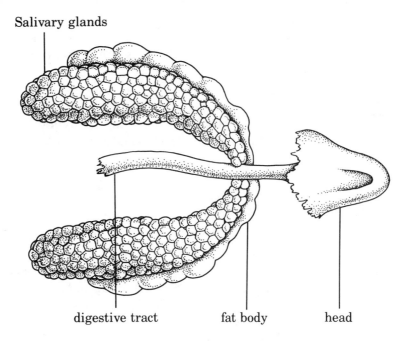

digestive tract   fat body   head

Fig. 17-3. Salivary glands of a *Drosophila.*

When you have found the glands, remove the extraneous material, and blot excess saline with a piece of paper towel, being careful not to soak up the glands as well. Add 2 drops of aceto-orcein stain, and leave the slide alone for ten minutes, being *sure* the stain does not dry out. Add another drop of stain during the staining period, if necessary. When staining is complete, remove most of the stain with a paper towel. Place a clean cover slip over the glands and a cork over the cover slip, and squash straight down with the cork. Pressure must be great enough to break the nucleus but not great enough to damage chromosomes. Examine the slide quickly, and, if it is worthwhile examining at greater length, seal the edges of the cover slip with Vaseline to prevent the slide from drying out.

Week 2   In this laboratory, you will prepare a second mating, using as parents the F_1 hybrids that have now emerged in the first culture. In addition, you will count and score (for eye color and sex) all the F_1 individuals.

Preparing a second mating

1. Prepare a culture vial, and label it appropriately.

2. Lightly etherize the F_1 flies until they have become motionless. *Do not kill them.* Select four males and four females, being certain of sex. *Do not use an animal if you are uncertain of its sex.* Put the flies into the new culture vial. Make sure that they have returned to normal (flying around, not caught in the medium) and that the vial is properly labeled before you return it to the supply desk.

3. Kill the remaining flies by overetherizing them, and classify them by sex and eye color. Record these data in Table 7. (*Important note:* If your original culture was started from red-eyed females and white-eyed males, and this week you find progeny with white eyes, there has been a mistake in the preparation of your original culture. If your original culture was the reciprocal and you find no white-eyed males, there was an error in its preparation. If either of the above applies to your culture, follow the remaining instructions, using flies from another student's culture.)

4. Thoroughly wash and rinse the old culture vial, and return it, upside down, along with its stopper, to the supply desk.

Week 3   Remove the F_1 hybrid parents, and discard them; make sure these parents are the same as indicated on the culture vial.

Determining the F_2 generation

Week 4   Classify the F_2 flies by sex and color, and record the data in Table 8.

1. Which of the alleles is dominant? How do you know this from your data?

2. Knowing which allele is dominant and knowing the nature of sex linkage, predict the genotype and phenotypic ratios for the F_1 and F_2.

3. Subject your data to chi-square analysis. Are the deviations statistically significant?

4. The instructor will lead a discussion of results and perform a chi-square analysis on data gathered from the entire class.

MULTIPLE ALLELISM

Many—indeed, most—gene loci in many species of organisms may be represented by more than two alleles. Generally, all the alleles in a given series affect the same character (quantitatively), but they may control different aspects of a character. Most genes also have **pleiotropic effects** (i.e., they affect more than one character). Where multiple alleles have recognizable pleiotropic effects, the effects of members of a series on different characters may or may not run parallel.

We shall consider only three alleles in an important series of multiple alleles in human beings that determines blood types in the **A-B-O** series. This case is a good example of how multiple alleles work and will give you some knowledge of an important character in human beings. By typing your own blood, you will learn how blood groups are determined and will thus be introduced to some concepts from the fields of immunology and serology. After you have determined your own blood type, the genetic basis for these blood types will be considered.

Blood typing   On their surface, the red blood cells (erythrocytes) of human beings may contain proteins called **antigens**. The red blood cells with type A blood have antigens A; those in type B blood have antigens B; the red blood cells in type O blood have neither A nor B antigens; and type AB erythrocytes contain both. Blood plasma contains other proteins called **antibodies**, designated anti-A and anti-B. A person with type A red blood cells has no antibodies in the blood plasma to antigen A (anti-A), but has antibodies to any antigens his or her own blood cells do not have. If blood from one individual is mixed with blood or plasma containing corresponding antibodies, the red blood cells clump together, or agglutinate, due to the adhesion of their antigens. When blood transfusions are to be given, then, it is always best to obtain a donor of the same blood type as the patient. If such a donor is not available, blood of another type may be used provided that the plasma of the patient is compatible with the red blood cells of the donor. Donor antibodies are diluted during the transfusion and will not cause serious difficulty unless the transfusion is extensive.

An individual's blood type could be determined by mixing a drop of his or her blood with sera from persons of known blood types. It is more convenient, however, to use commercially prepared sera.

1. Using a heavy wax pencil, divide a perfectly clean slide into three compartments by drawing heavy lines and labeling the compartments as indicated:

A	B	CONTROL

2. Add a drop of serum containing anti-A antibody to the first compartment on your slide and a drop of serum containing anti-B antibody to the second compartment.

3. Puncture your finger with one of the sterile disposable lancets provided. *Do not use a lancet previously used by another student.* Add a drop of blood to each compartment on your slide, being careful to allow the blood to fall into the drops of sera or to transfer it with a *clean* toothpick, so

that there is no chance of transferring any of the sera from one compartment to another. Mix the material in the first compartment with the tip of a clean toothpick, and then discard the toothpick. With another clean toothpick, mix the material in the second compartment, and discard the toothpick. Stir the third drop with a third toothpick, and allow it to stand as a control. What is the value of a control in this experiment? (*Caution:* Do not use a single toothpick for more than one antiserum. Why?)

4. After about a minute, observe each drop on your slide. If agglutination has occurred, the erythrocytes will be seen sticking together in small clumps. If no agglutination has occurred, the erythrocytes will be scattered uniformly throughout the liquid. (Changes in appearance after two or more minutes indicate normal clotting and should not be regarded as agglutination. If you are uncertain of your results, use your dissecting microscope.) Which condition should be observed in the control? Record your findings in the table that follows, using the symbols given.

A	B	CONTROL

+ = agglutination
– = no agglutination

5. The left side of Table 9, below, shows the antigen and antibody content of blood of each of the four types in the A-B-O series. From these data you should be able to determine which types will give positive results when tested with the two types of sera; place plus or minus signs in the appropriate places in the two columns on the right side of the table.

GROUP	BLOOD CONTAINS		REACTION WITH ANTIBODIES	
	CELLULAR ANTIGENS	PLASMA ANTIBODIES	ANTI-A	ANTI-B
O	None	anti-A and anti-B		
A	A	anti-B		
B	B	anti-A		
AB	A and B	None		

TABLE 9.

6. Using the information in Table 9, determine the answers to the following questions:

a. What is your blood group?

b. What type (or types) of blood could be safely transfused into your circulation?

c. Into what type (or types) of recipient(s) could your blood be safely transfused?

Explain your answers to b and c, above.

Genetics of the A-B-O Blood Groups The inheritance of the A-B-O groups is best described by a theory of three alleles, which are designated I^A, I^B, and i (or sometimes L^A, L^B, and L^O). Both I^A and I^B are dominant over i, but neither I^A nor I^B is dominant over the other when both are present in the heterozygote. Accordingly, the four blood groups, or phenotypes, correspond to the genotypes indicated below:

BLOOD GROUP	GENOTYPE
O	i/i
A	I^A/I^A or I^A/i
B	I^B/I^B or I^B/i
AB	I^A/I^B

On this basis, what is your genotype? _____ If you know the blood groups to which your siblings or parents belong, you may be able to be more specific.

Table 10 gives the approximate frequency of each phenotype in the United States population. The frequencies are often different in the populations of different nationalities and geographic regions, and, as a result, can be used in studying the origins and relationships of different peoples. As you can readily guess, there is much variation in the United States population because of its diverse origins. In the table, record the blood-group frequencies for the entire class. Using a chi-square test, compare your results with those indicated.

FREQUENCY IN POPULATION		PHENOTYPE	NUMBER	FREQUENCY
(U.S. WHITE)	(U.S. BLACK)	GROUP	IN CLASS	IN CLASS (%)
45%	47%	O		
41%	28%	A		
10%	20%	B		
4%	5%	AB		
100%	100%		TOTAL	100%

TABLE 10.

In our judicial system, blood typing is often used as evidence in paternity cases. In a series of disputed paternity cases, the mother and child in each had the blood types listed in Table 11. For each, indicate the blood types which, if found, would exonerate an accused man.

Determining blood type in disputed paternity cases

MOTHER	CHILD	MAN EXONERATED IF HE BELONGS TO GROUP
A	O	
B	AB	
O	A	
AB	A	
O	O	
B	B	
A	B	
AB	AB	
A	A	
A	AB	
B	A	
B	O	
AB	B	

TABLE 11.

MULTIPLE GENE COMPLEXES

Often, several different genes affect the same character. The ways in which such genes interact vary greatly. Some have additive effects, some are antagonists, etc. It is also common for several independent genes to affect closely related but different characters; the characters then interact in their effect on the life of the organism. Here you will study an example of the latter type of genetic complex.

You learned earlier that the sense of taste in human beings is commonly analyzed in terms of four basic tastes—sweet, sour, bitter, and salty. The sensitivity to each of these tastes is controlled by many different genes. Thus, one gene may determine whether substance A is tasted as sweet, a second gene may determine whether substance B is tasted as sweet, a third may do the same for substance C, etc. Each of these genes controls a separate character, but the characters all interact to affect the way in which a person tastes a meal. You will test yourself for several such genetic taste factors in this section.

Testing for genetic taste factors

1. Obtain a piece of test paper that has been treated with the chemical *p*henyl*t*hio*c*arbamide (PTC). Making sure that your tongue is moist, touch the paper lightly to your tongue, and then record whether or not you taste the PTC. If the reaction is positive, record whether the sensation is a sweet, sour, bitter, or salty one. If you did not taste the PTC, try the experiment again by chewing a small piece of the paper, recording results as above.

The ability to taste PTC is inherited and is apparently determined by a pair of alleles. Normally, the allele for tasting is dominant over that for nontasting. In the United States, about 70% of the people are tasters (the great majority taste the chemical as bitter), while about 30% are nontasters. The percentages vary greatly among different human stocks and are used in research to study various populations, as are those of blood types.

It should be emphasized that the ability to taste PTC is not always clear-cut; there is much variation in the sensitivity threshold. Also, it is interesting to note that the chemical can be tasted only if it is dissolved in the taster's own saliva; even a sensitive taster cannot detect the PTC if it is dissolved in water or in someone else's saliva and then placed on his or her dry tongue.

2. Obtain a piece of thiourea test paper, and, following the procedures used above, determine whether you are a taster or nontaster. This substance is closely related chemically to PTC, but the ability to taste it is inherited independently. Thus, although most people can taste thiourea (as in the case of PTC), the taster and nontaster groups for the two substances need not be the same.

3. Obtain a piece of sodium-benzoate test paper, and determine your ability to taste it. If you are a taster, determine whether the sensation is a sweet, sour, bitter, or salty one. The ability to taste sodium benzoate is inherited independently of the sensitivity to PTC, but the two taste characters apparently interact markedly in their effect on a person's reaction to various foods. For example, people who taste PTC as bitter and sodium benzoate as salty tend to like such foods as sauerkraut, buttermilk, turnips, and spinach more than the average person, while people who fall into the bitter-bitter group for the two test chemicals like those foods less than the average person. Giving PTC first and sodium benzoate second, the most common taste group seems to be bitter-salty, followed by bitter-sweet, bitter-bitter, and tasteless-salty. Most other combinations also occur.

QUESTIONS FOR FURTHER THOUGHT

1. Construct a diagram showing independent assortment of alleles during meiosis.

2. How does meiosis in sperm formation differ from that in egg formation?

3. What steps would a plant breeder take to obtain seeds like those you used in your tobacco-seedling experiments? How could the user of such seeds be absolutely certain that the seeds represented genuine offspring from a monohybrid cross? How could a 3:1 ratio be "faked"?

4. What is a sex-linked allele?

5. What kinds of characteristics make an organism suitable for breeding experiments? What groups of organisms would make especially useful subjects for such experiments? Would human beings be useful for genetic studies?

6. In one sentence, carefully explain the significance of the chi-square test. What does "statistically significant" mean?

7. Type A blood contains both antigen A and anti-B antibody. Type O blood contains neither antigen but both anti-A and anti-B antibodies. Why is it safe to transfuse moderate amounts of type O blood into an individual whose blood type is A?

TOPIC 18 HUMAN
KARYOTYPES

Biological Science, 4th ed.
pp. 619, 675–680, 690
Elements of Biological Science, 3rd ed.
pp. 391–393, 424–425

Every organism has a characteristic number of chromosomes in each of its cells. In humans, the normal chromosome complement is 46 chromosomes. This chromosome number was not correctly established until 1957; previously, it was thought that the human genome contained 48 chromosomes. The human genome is composed of 22 pairs of **autosomes** and 1 pair of **sex chromosomes** designated X and Y. Normal females have 2 X chromosomes, while the male has an X chromosome and a Y chromosome. During oogenesis all eggs that form receive 22 of the autosomes and an X chromosome, while in spermatogenesis half of the sperm cells receive an X chromosome plus 22 autosomes and half receive the Y chromosome plus 22 autosomes. If chromatids fail to separate during meiosis, some gametes will contain extra chromosomes while other gametes will be lacking genetic material. When chromatids do not separate normally during meiosis, this is referred to as **nondisjunction**. If one of these gametes with extra genetic material combines with a gamete with the normal haploid chromosome complement, the result is an individual with an extra chromosome.

An important development in the study of human chromosomes was the advent of new staining techniques resulting in the appearance of characteristic banding patterns on the chromosomes. When chromosomes are arrested in metaphase, each chromosome condenses into a characteristic size and shape, and exhibits a specific banding pattern when stained. These bands represent differences in the chemical composition and organization of the chromatin. The exact banding pattern in chromosomes has remained stable over long periods of evolutionary time and suggests that the organization of the chromatin is extremely important to proper gene function. From the study of various banding patterns, linkage analysis, and cell hybridization, cytogeneticists have been able to identify over 1,500 loci on the autosomes and over 100 loci on the X chromosome. Because of the small size of the Y chromosome, only a few genes are located on that chromosome.

Chromosome squashes can be prepared from white blood cells or from cells from the amniotic fluid. Amniotic fluid is used to detect abnormalities in the developing fetus. After the cells have been cultured and stopped at metaphase, they are stained and photographed, and a karyotype can be prepared. A **karyotype** is the arrangement of paired chromosomes in decreasing size, and identified according to standard nomenclature.

Figure 18-1 shows a stylized banded karyotype (keep in mind that there are two of each of the chromosomes in metaphase), while Figure 18-2 shows an actual karyotype of a normal female. Each chromosome has 2 **chromatids**, joined at the **centromere**, which is the site of attachment to the spindle fiber. The position of the centromere is constant for each chromosome, resulting in a short and long arm. The ratio of these lengths are used to identify the

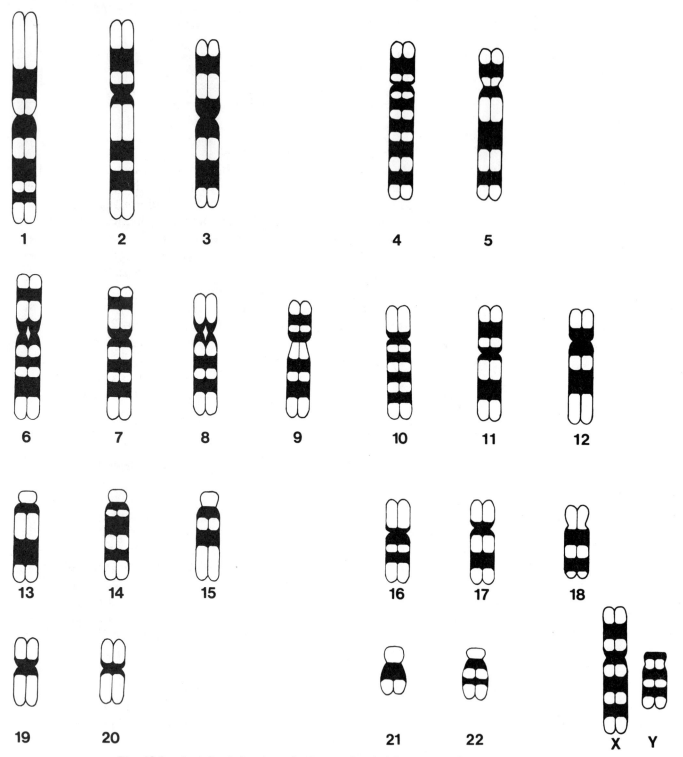

Fig. 18-1. A stylized drawing of a human banded karyotype. *(Courtesy Philip Harris Biological Ltd.)*

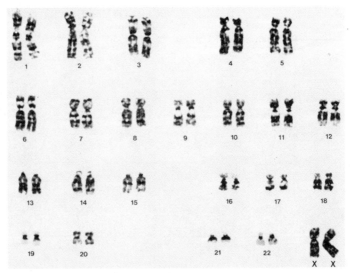

Fig. 18-2. Karyotype of a normal female. *(Courtesy Philip Harris Biological Ltd.)*

number to which these chromosomes are assigned. Note that the sex chromosomes are not assigned a number.

Determining phenotype

In this exercise, your instructor will give you a chromosome squash of either a normal individual or of an individual with one of the abnormalities listed below. Count the number of chromosomes, and record the number on the karyotype chart provided. Cut out the chromosomes, place them on double-stick tape, and arrange them according to size and banding pattern on the karyotype sheet. Determine the phenotype of the individual with this kind of genotype. If this genome contains an abnormal chromosome pattern, explain how this chromosome abnormality occurred.

CHROMOSOMAL SYNDROMES

Cri du chat (Cat-Cry) Syndrome   This syndrome was first identified in 1963. A child born with this syndrome has the normal chromosome number; however, the terminal portion of the short arm on chromosome 5 is deleted. The name of the syndrome comes from the catlike cry characteristic of babies with this disorder. Other features accompanying this disorder are severe mental retardation and cardiovascular disease.

Down's Syndrome (Trisomy Syndrome)   This syndrome is the result of an extra number 21 chromosome. This is the most common cause of malformed humans, occurring once in every 500 to 1,000 births. A person with Down's syndrome generally has a low IQ, and sexual development is poor. Almost all of the body tissues may be affected in some manner. A definite relationship has been established between women over thirty-five years of age giving birth and the occurrence of this syndrome.

KARYOTYPE CHART

CHROMOSOME NUMBER _____   PHENOTYPE_____

1   2   3   4   5

6   7   8   9   10   11   12

13   14   15   16   17   18

19   20   21   22   Sex
chromosomes

Klinefelter's Syndrome   Individuals with this syndrome have 47 chromosomes, including XXY sex chromosomes. Approximately 1 male in 500 to 800 are affected with this abnormality. Males with Klinefelter's have poor gonadal development and may have mild mental retardation but a normal life span.

Male iso-21 Syndrome   In this syndrome, the male has the normal chromosome number, but chromosome 21 has two identical long arms instead of a normal chromosome 21. This type of abnormality develops when the centromere divides horizontally rather than longitudinally during meiosis (Figure 18-3). The small chromosome is lost. The individual carrying an iso-21 and a normal 21 chromosome has characteristics similar to individuals with Down's syndrome.

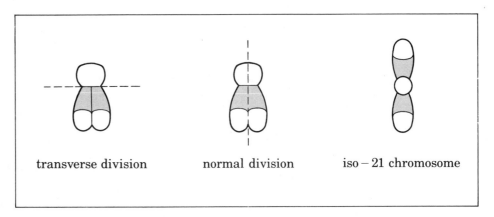

transverse division   normal division   iso – 21 chromosome

Fig. 18-3.   Isochromosome formation.

XYY Syndrome   Individuals with this syndrome have 47 chromosomes, are frequently tall, and may have lower-than-normal intelligence. The frequency of this syndrome is about 1 in 1,000 males. This syndrome has generated controversy. Since it was observed that a higher percentage of males in mental and penal institutions have this genotype as compared to males in the general population, researchers have suggested that this genome predisposes an individual to aggressive behavior. This, however, has not been proved.

QUESTIONS FOR FURTHER THOUGHT

1. Why is meiosis important in gametogenesis?
2. How do oogenesis and spermatogenesis differ?
3. During spermatogenesis, compare the type of sperm formed when nondisjunction occurs in meiosis I versus meiosis II.
4. What would happen if nondisjunction occurs in the zygote?

The flowering plant is far from a simple colony of cells; it is a complexly organized unit whose growth, development, and behavior are precisely regulated and synchronized. The plant's synchronous growth and development as well as its ability to respond adaptively to environmental fluctuations are achieved largely through the action of chemical agents called **plant hormones,** substances generally produced in the apical meristems of the root and shoot and then transported to other parts of the plant by the vascular system or by diffusion.

In this unit, you will explore some of the normal events in plant growth and development. In the following unit, we shall study experimentally some of the hormonal controls that natural selection has shaped to the plant's advantage as regulators of behavior and development.

GAMETE DEVELOPMENT

Male gametophyte

The formation of the male gametes occurs within the anther of the flower. Initially, the anther begins as a mass of undifferentiated cells, but eventually four distinct groups of cells develop which will form the four male **sporangia** (Figure 19-1), or pollen sacs. Some of the cells differentiate into the walls of

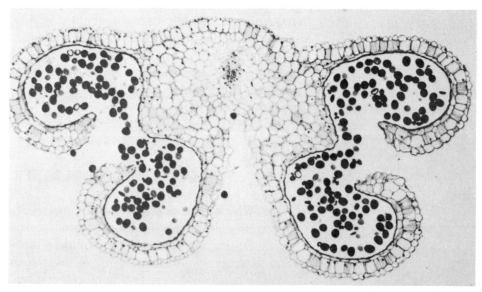

Fig. 19-1. Lily anther bearing four sporangia with pollen. *(Courtesy Carolina Biological Supply Company.)*

the sporangia, and some into the nutritive cells for the developing gameto- phyte. Each of the cells in the central portion of the sporangia undergoes meiosis and gives rise to a **tetrad** or group of four haploid **microspores**. In most angiosperms, the microspores separate, each developing a thickened wall that will protect the male gametophyte until germination on the stigma. The formation of the male gametophyte or pollen grain is complete when the microspore divides by mitosis, giving rise to a **generative nucleus** and **tube nucleus**. In many plants, the pollen grain is released from the sporangi- um in the two-celled stage, while in others, the generative nucleus divides to form the two **sperm cells** which are the male gametes.

1. Obtain a flower, and remove an anther. Make a wet mount of a cross section through the anther. Locate the sporangia and pollen grains. In mature anthers, the pollen sacs split open, releasing the pollen.

2. Examine prepared slides of lily anther showing pollen tetrads and mature pollen grains. Note the two distinct nuclei in each of the pollen grains. What is the function of the generative nucleus?

Female gametophyte   The development of the female gametophyte occurs inside the ovary in the **ovule** or sporangium. Early in the development of the ovule, a single cell, larger than the surrounding cells, begins to differentiate. This cell will give rise, through meiosis, to four haploid **megaspores**. Three of the megaspores usually disintegrate, and the fourth develops into the female gametophyte.

The remaining haploid megaspore divides mitotically three times, pro- ducing eight nuclei. Two of the eight nuclei migrate to the center of the cell and form the **polar nuclei**. One of the other nuclei will develop into the **egg cell**. Cell walls develop around the nuclei, resulting in an eight-nucleate, seven-cell female gametophyte, also referred to as the **embryo sac** (Figure 19-2).

Observe the prepared slides, showing the stages of female gametophyte development, which are on demonstration.

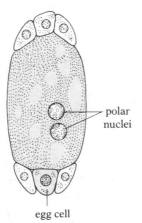

polar nuclei

egg cell

Fig. 19-2. Female gametophyte (embryo sac), which is composed of seven cells. One cell is much larger than the others and contains the two polar nuclei.

EMBRYONIC DEVELOPMENT

Like most other multicellular organisms, a flowering plant begins life as a **zygote**, which results from the fertilization of the egg by one of the sperm nuclei. In the ovary of the parent plant, the zygote develops into a small **embryo**. The food for this growing embryo is provided by the endosperm tissue, which develops from the fusion of the second sperm nucleus and the polar nuclei. Together with food materials and adjacent ovarian tissue, the embryo is formed into a **seed**. The ovary, which may contain many seeds, develops (sometimes in association with other ovaries in the same flower) into a **fruit**. Both seed and fruit, as we shall later see in more detail, are important adaptations that help ensure the success of terrestrial reproduction.

The embryo often enters a period of dormancy before resuming growth (**germination**) under the proper conditions of light, moisture, temperature, and oxygen supply. When germination occurs, additional cells are produced in meristematic zones; the growth and differentiation of these cells, and their organization into plant organs—roots, stems, leaves, flowers, etc.—result in a structural pattern characteristic of the species.

Examining elements in germination

1. Examine a soaked germinating bean seed. Locate and carefully remove the **seed coat**. This tissue protects the embryo and is often modified to absorb water or to facilitate dispersal of the seed by air, animals, or other agents. The fruit in which the seeds are borne may be similarly modified. On the bean seed, note the small, rounded **hilus** in the shallow concavity on one side of the seed. This structure serves to absorb water, while the rest of the seed coat is impermeable to water. Some seeds have a very hard water-impermeable seed coat that must be broken or abraded before water absorption can occur. What value might there be in such a seed coat?

2. Split the seed open, and locate the two large **cotyledons**, leaves modified for food storage. The cotyledons are part of the embryo, and the food they store is principally in the form of starch. The presence of two cotyledons is characteristic of dicots (subclass Dicotyledoneae of the flowering plants—see Appendix 2). What are some of the other differences that distinguish monocots from dicots?

3. The embryonic plant axis is located between the cotyledons. The region sheathed at its tip in embryonic leaves is the **epicotyl**, and the opposite end of the axis, on the other side of the cotyledonary attachment, is the **hypocotyl**. Each terminates in an apical meristem. The epicotyl and part of the hypocotyl form the shoot (stem) of the plant; the remainder of the hypocotyl as well as the radicle form the root. Notice the position of the cotyledons on the maturing bean plants, on demonstration.

The original food in the seed was **endosperm**, a tissue derived by cell division and growth from the same cell that gave rise to the egg; during development of the bean embryo, its cotyledons absorbed all the endosperm. If a castor-bean seed is available, dissect it carefully and note the thin, veined cotyledon. The rest of the interior of the seed, apart from the embryo, is endosperm. *Do not eat the castor-bean seed;* it is poisonous.

GROWTH AND DIFFERENTIATION OF THE ROOT AND SHOOT

Primary growth, as you recall from Topic 7, is the result of cell division in *apical* meristems and of differentiation in the cells thus produced. *Secondary* growth is the result of cell division in *lateral* meristems (vascular cambium and/or cork cambium) and the differentiation of these cells into xylem, phloem, or cork. Primary growth increases the length of an organ, while secondary growth results in increase in girth. We shall investigate the zones of primary growth in a root and some of the patterns to which root growth gives rise; we shall then turn our attention to shoot development.

The locus of primary growth in a root

Root Development and Differentiation The pattern of primary growth in a root may easily be studied by marking the root at known regular intervals and then comparing the position of the markings with the original positions after twenty-four to forty-eight hours. *Read the directions carefully,* and obtain all the necessary equipment before proceeding.

1. Obtain a 3 × 3–inch piece of thin glass or Plexiglas and 4 germinated corn or pea seedlings, each with straight root about 2 cm long. Cover the glass plate with wet paper toweling cut to fit, and fasten the seedlings over the toweling with a rubber band, as shown in Figure 19-3. Being sure *not* to let the seedlings dry, mark them as directed below.

2. Blot any excess moisture from the seedlings. With a marker like that shown in Figure 19-3 (made from a paper clip or soft wire and thread), mark the root with India ink nine times at 1-mm intervals, *starting at the root tip.* Paint the thread with India ink or other black drawing ink, using the ink applicator or a dropper. Wipe excess ink from the thread, and quickly make a thin mark on the root. You can make the marks easily if a millimeter ruler is slipped under the root during the marking operation. (*Note:* Be sure to wipe excess ink from the thread. If you do not, the marks will later smudge and be difficult to interpret.)

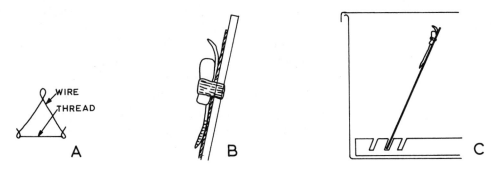

Fig. 19-3. Root-growth experiment. A: Marker; B: diagram showing the seedling attached to the plate with a rubber band; C: plate in a holder, in a moist chamber.

3. With a solvent marking pencil, mark the glass on the back with your name or seat number. Place the whole setup in a moist chamber at the supply desk, slipping it into one of the 60°-angled slots in the holder on the

DISTANCE BETWEEN MARKINGS = 1mm	ROOT NO.	DISTANCE BETWEEN MARKINGS									FINAL SKETCH (SCALE =)
		TIP-1	1-2	2-3	3-4	4-5	5-6	6-7	7-8	8-9	
	1										
	2										
	3										
	4										
INITIAL SKETCH (SCALE =)	AVG.										

TABLE 1. THE LOCUS OF GROWTH IN A ROOT.

chamber bottom. In Table 1, make a scale sketch of the root, several times enlarged, showing the position of the markings and the length of the root relative to the size of the seed.

4. After forty-eight hours (or, alternatively, using another student's specimen set up forty-eight hours ago), sketch the root again, to the same scale used for the previous sketch. Measure the distance between the markings, record them in Table 1, and use the averages to make markings on the new sketch. Where did the most elongation occur? How far behind the root tip has elongation stopped taking place? If your initial lines were neat and clean, how do you account for smudging that has taken place a short distance behind the root tip?

Would a control have enhanced the value of this experiment? What would or could be used as a control?

Pattern and lateral root formation

As we shall see shortly, branches on a stem form from specialized meristems called lateral buds. Formation of lateral roots is much more random, since the laterals form through meristematic activity of the *pericycle,* one of the cellular layers in the root adjacent to the vascular tissues. *Secondary* roots formed in this way may repeatedly rebranch by the same method. Examine a demonstration slide showing formation of a lateral root. Identify the pericycle and the lateral root.

What adaptive advantage might there be in greater flexibility in the pattern of branching in the root as compared to the regular pattern of branching in a stem? What advantages would there be in regular branching of the stem?

Stem development and differentiation

(*Note:* Unless the weather is inclement, most of this study can be conducted out of doors. Most of the following observations can be made on bare over-wintering tree twigs or on conifers.)

Stems, like roots, grow in length from apical meristems and in girth either by cell enlargement (most monocots) or by growth from lateral meristems (like the vascular cambium and cork cambium in many dicots).

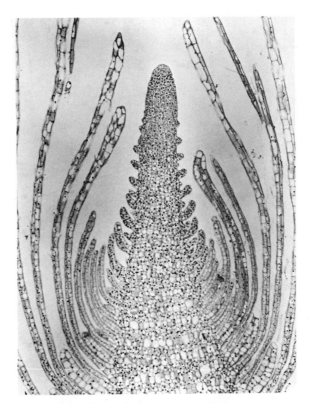

Fig. 19-4. Longitudinal section through the apical meristem of *Elodea.* *(Courtesy Thomas Eisner, Cornell University.)*

Figure 19-4 shows the apical meristem of *Elodea,* a water plant, sheathed by developing young leaves and by **leaf primordia,** budlike structures that will give rise to leaves. Notice that the primordia become smaller as they near the stem tip. Differentiation in stems, like that in roots, proceeds as a region grows farther from the center of meristematic activity that produced it.

1. Apical meristems are of two kinds on most stems; one or more **terminal buds** may occur at the tip of a shoot, and **lateral (axillary) buds** occur in the upper angle between leaf petioles and stem axis. (If there is no foliage on the plant you are examining, locate buds adjacent to **leaf scars,** regions where leaves were previously attached.) In a temperate perennial plant such as a tree, both types of bud enter a period of dormancy in the summer or fall and renew their growth the following spring. Examine a twig several years old that shows both kinds of bud.

2. Each season's growth is delimited by a group of several **bud scale scars,** tissue that remains after the scales covering each bud fall off in the spring. Like annual rings, bud scale scars are indicators of age in stems. How many years' growth are represented by the section of stem you are examining?

The presence of lateral buds is diagnostic of stems. Look, for example, at a compound leaf on demonstration. This leaf could be mistaken, on superficial examination, for a section of leafy stem, but notice the absence of buds. Growth of a lateral bud produces a *branch,* a new section of stem with

Fig. 19-5. Portion of a red maple twig. *(Courtesy Carolina Biological Supply Company.)*

an entirely new set of meristems. Whether a lateral bud develops depends upon its proximity to the apical meristems of other branches and of the main shoot. Hormones transported from active apical meristems inhibit development of lateral buds, but this influence diminishes with distance from the active meristems.

Note that leaf scars are clearly evident only in the youngest sections of most twigs for the same reason that bud scale scars are useful indicators of age only in young sections of a woody stem. Label all parts of the maple twigs shown in Figure 19-5. How many years' growth is represented by these twigs?

3. Notice on a deciduous tree or on a branch provided by the instructor that growth of leaves is usually restricted to twigs formed in the current growing season. If you are examining leafless trees, all next year's leaves will be formed on the sections of stem axis that emerge from terminal and lateral buds on last year's twigs. Conifers and other evergreens are exceptions to this rule since although their leaves (needles) fall, they do not fall all at once, as is the case in deciduous trees; leaves of more than one growing season may be present simultaneously. Considering the geographical distribution of conifers and deciduous trees, what do you think might be the significance of the differences between the leaf-form and leaf-fall patterns of these two types of trees?

4. Examine demonstrations of modifications in stems and leaves. Each one will be accompanied by an explanatory card. Of what advantage would each of these adaptations be?

TOPIC 20

HORMONAL CONTROL OF PLANT DEVELOPMENT

Biological Science, 4th ed.
pp. 378–385, 390–391
Elements of Biological Science, 3rd ed.
pp. 237–242

Though we commonly associate the term "behavior" with the overt activities of animals, we are constantly reminded that plants can respond with remarkable versatility to environmental change. Plant stems, for example, turn actively toward light; roots turn toward gravity. Leaf petioles bend and turn, exposing the leaf blades to sunlight at the most effective angles. Most such adaptive movements are mediated by hormones. Plant hormones play an important governing role in a wide variety of other integrative and developmental processes such as the origin, aging, and abscission (dropping) of leaves, flowers, and fruits; the development of species-specific patterns; and the appearance and course of cellular differentiations in growing regions of the plant.

Some of the primary effects of hormone action are well known. Hormones may stimulate or inhibit cell enlargement or cell division with consequent changes in the shape, size, or orientation of the plant part in question; hormones may stimulate or inhibit cellular differentiations. Often several hormones interact complexly to produce effects different from those that would be produced by the activity of individual hormones. The precise action of a plant hormone depends on the hormone in question, its concentration, and the nature of the responding tissue. The mechanisms by which hormones act are poorly understood, although they are presently being actively researched.

Apart from their inherent interest and importance as the principal regulators of plant growth, hormones find a wide variety of commercially valuable applications. By using the appropriate hormone, for example, one can induce or break dormancy in some seeds or tubers, increase the tenderness of celery stalks by accelerating their growth, or produce seedless fruits. A knowledge of how hormones act differentially on different species of plants makes possible increasingly effective weed control, and an understanding of how hormones function to influence plant pattern often makes possible the control of pattern to suit agricultural or horticultural requirements.

In this unit, you will work experimentally with the chemical control of plant growth, development, and pattern.

(*Note to the instructor:* Topic 20 is for students who will be able to check regularly on experimental plants kept in the laboratory and in the greenhouse. If laboratory logistics prohibit students' regular attention to their experiments or if greenhouse facilities are not available, please refer to the *Instructor's Guide* for suggested modifications.)

THE INFLUENCE OF AUXINS ON STEM AND ROOT GROWTH

The discovery of naturally occurring plant growth regulators dates largely from the experiments of Charles Darwin, who showed that the tip of the coleoptile (sheath) covering the young shoot of an oat plant was necessary for a phototropic response. It was eventually discovered that this response involved a chemical substance manufactured in and transported from the plant tip to other parts of the plant. This substance was named **auxin**; later, it was isolated and found to be indoleacetic acid **(IAA)**. For many years, auxin was synonymous with IAA; it has been found, however, that a variety of both naturally occurring and synthetic compounds have similar effects. Many such compounds have elements of chemical structure in common with IAA.

IAA and other auxins act by stimulating cell elongation; they affect the uptake of water and the elasticity of the cell wall. By some type of active transport that is not understood, they accumulate on the shaded side of a differentially illuminated shoot, bringing about a more rapid elongation of cells on that side of the shoot than on the other and hence a bending of the shoot toward a light source. If the auxin source (the shoot tip) is excised and then replaced off-center or if the tip is placed on a small gelatin block which is then placed off-center on the shoot, this response will occur even in the dark (Figure 20-1). In each case, cell elongation occurs on the side of the shoot where the auxin source was replaced.

The placing of an auxin-containing gelatin block on the cut tip of the coleoptile is the basis for an auxin-activity test called the *Went growth curvature test*. Any substance showing coleoptile-bending activity is classified as an auxin. Because the concentration-response relationship is well known and because the response is so sensitive to small differences in auxin concentration, the growth curvature test is commonly used as an indicator, or **bioassay**, of auxin concentration. This general principle—use of an organism to indicate the presence or concentration of a substance in amounts too small to

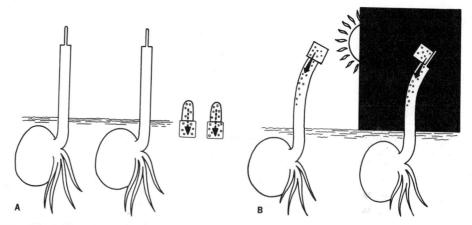

Fig. 20-1. Went's experiment. When the tips of coleoptiles are cut off and placed on blocks of agar for about an hour (A), and one of the blocks alone is then put on a stump (B, left), the stump will resume growing and will respond to light. If a block is placed off-center on a stump in the dark (B, right), the stump will grow and will bend away from the side on which the block rests. Apparently a hormone has diffused from the tips into the blocks, and this hormone can then diffuse from the blocks into the stumps.

Fig. 20-2. Razor-blade cutter for making sections of uniform size.

be detected by conventional microanalysis—finds application in many fields of biology.

Because success of the curvature test requires rigidly controlled environmental conditions, we shall illustrate bioassay by using a similar but not so sensitive test: the *straight-growth test.* In addition, we shall see that shoots and roots respond quite differently to IAA.

The straight-growth test

In this test, straight sections of coleoptile or stem of known length are placed in auxin solutions of unknown concentration. After a specified period of time, the sections are removed and their new lengths measured. To determine the unknown concentrations, these data are then compared with known information about the relationship between auxin concentration and growth for the plant in question.

1. Working individually, or in groups as directed by your instructor, obtain eight petri dishes. Label six dishes with the numbers 3, 4, 5, 6, 7, and 8, to correspond to 10^{-3}, 10^{-4}, 10^{-5}, 10^{-6}, 10^{-7}, 10^{-8} M concentrations of IAA. On the seventh dish, write "unknown"; on the eighth, write "control." Add your name to each dish top. Obtain the appropriate dilutions of IAA, the unknown solution, and the control solution (one containing no IAA) from your instructor; half-fill each petri dish with the solution indicated on its label.

2. Obtain several etiolated (dark-grown) shoots of Alaska pea, and place them on a wet paper towel. Carry out all subsequent handling on the towel, and enclose the shoots in it to prevent desiccation when they are not actually being handled. Using the apparatus illustrated in Figure 20-2, cut 24 5-mm sections from the *youngest internode* (that straight region just below the terminal curvature, as shown in Figure 20-3). As the sections are cut, put them quickly into the petri dishes until you have three such sections in each dish. Cover the dishes, put them in a dark place, and do not disturb them for twelve to twenty-four hours. In Table 1, record the initial section lengths, which should all have been the same.

3. At the end of the incubation period, record in mm the final length for each section, and enter the data in Table 1. Compute the *average* elongation for each group, and plot these averages against IAA concentration on Graph A.

Fig. 20-3. Dark-grown pea shoot.

Which concentration of IAA was most effective in promoting elongation of the sections? Is there any evidence that higher concentrations are inhibitory?

What is the approximate concentration of the unknown IAA solution?

Why is the "control" dish necessary? Did the stem sections in it show any growth? How did the control differ from the experimentals?

Root-growth test   1. Obtain eight petri dishes. Label six dishes with the numbers 5, 6, 7, 8, 9, and 10, to correspond to 10^{-5}, 10^{-6}, 10^{-7}, 10^{-8}, 10^{-9}, 10^{-10} M concentrations of IAA. Label the seventh dish "unknown," the eighth "control," and put your name on all eight dishes. Place a disc of filter paper in each dish,

A	INITIAL				B	FINAL (AFTER HRS)				AVERAGE ELONGATION
	1	2	3	AVG.		1	2	3	AVG.	
10^{-3} IAA					10^{-3} IAA					
10^{-4} IAA					10^{-4} IAA					
10^{-5} IAA					10^{-5} IAA					
10^{-6} IAA					10^{-6} IAA					
10^{-7} IAA					10^{-7} IAA					
10^{-8} IAA					10^{-8} IAA					
UNKNOWN					UNKNOWN					
CONTROL					CONTROL					

TABLE 1. EFFECT OF IAA CONCENTRATION ON SHOOTS.

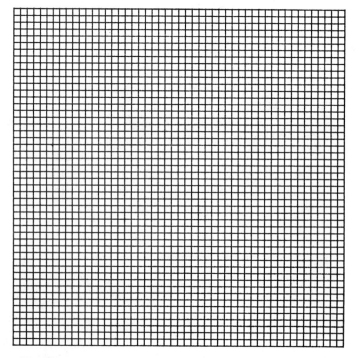

GRAPH A.

and soak the paper with 5 ml of the appropriate IAA dilution, unknown, or control (no IAA) solution.

    2. Obtain 24 healthy germinating seedlings of Alaska pea with roots of approximately the same length (about 1 cm). Keep the plants wrapped in a wet paper towel except when actually manipulating them. Measure the root length of three of the seedlings, and record the data in the first block of Table 2. Average these measurements. Place the three seedlings into the dish numbered 5.

    3. Repeat this procedure for the other five concentrations and for the unknown and the control.

A	INITIAL				B	FINAL (AFTER HRS.)				AVERAGE ELONGATION
	1	2	3	AVG.		1	2	3	AVG.	
10^{-5} IAA					10^{-5} IAA					
10^{-6} IAA					10^{-6} IAA					
10^{-7} IAA					10^{-7} IAA					
10^{-8} IAA					10^{-8} IAA					
10^{-9} IAA					10^{-9} IAA					
10^{-10} IAA					10^{-10} IAA					
UNKNOWN					UNKNOWN					
CONTROL					CONTROL					

TABLE 2. EFFECT OF IAA CONCENTRATION ON ROOTS.

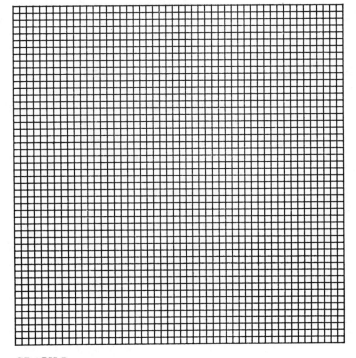

GRAPH B.

4. After twenty-four to forty-eight hours, remeasure and record the three primary root lengths (ignore any lateral roots) in each concentration set, and average them. On Graph B, plot your data as for the straight-growth test, with root elongation on one axis and IAA concentration on the other.

In which of the solutions did the greatest increase in root length occur?

How would you describe the reaction of the roots to this range of IAA concentrations?

Which of the two tests (shoot elongation or root elongation) represents a more sensitive assay of IAA concentration? Be prepared to defend your answer.

AUXIN-INDUCED TROPISMS

The substantial difference in response of roots and shoots to IAA reflects a differential sensitivity to auxin that natural selection has shaped to the plant's advantage in an important behavioral response—**geotropism**. Roots exhibit *positive* geotropism—i.e., they grow toward gravity. Since seeds are often scattered randomly in the soil and, therefore, do not always germinate with the root end downward, positive geotropism is obviously an adaptive response. Stems exhibit a *negative* geotropic response, which is similarly adaptive, since it tends to push them into the light above the soil.

Both responses can be explained on the basis of auxin distribution in the organ concerned. In the absence of other stimuli such as light, and if the organ is horizontal, auxin produced by either shoot or root meristem tends to accumulate in the lower portion of the organ near an active meristem.

In the stem, with its relatively high threshold of auxin sensitivity, this stimulates a greater elongation in the lower cells and thereby brings about an upward bending. In the root, with a much lower threshold of auxin sensitivity, a similar auxin concentration *inhibits* elongation on the lower side of the root so that more elongation occurs on top; bending is, therefore, in a downward direction.

Root geotropism

1. Obtain a 250-ml beaker from the supply desk, and line it with several thicknesses of wet paper toweling. Fill the beaker loosely with moistened sphagnum or vermiculite, and obtain half a petri dish or an appropriately trimmed piece of Parafilm to serve as a cover. From the germination dishes at the supply desk, obtain 4 corn seedlings, each with a root about 1 cm long. Wrap them in a wet paper towel, and keep them wrapped except when you are manipulating them for the experiment. The seedlings must not be allowed to dry.

2. Measure to the nearest mm and record in Table 3 the length of the root of 1 seedling. Place this seedling between the toweling and the glass wall of the beaker, root oriented directly downward. With a wax marking pencil, label the outside of the container "A," adjacent to the seedling.

3. Repeat this procedure with the 3 other seedlings, measuring each

INITIAL	AFTER 12 HOURS	AFTER 24 HOURS	AFTER 48 HOURS
A			
LENGTH(mm)			LENGTH(mm)

INITIAL	AFTER 12 HOURS	AFTER 24 HOURS	AFTER 48 HOURS
B			
LENGTH(mm)			LENGTH(mm)

INITIAL	AFTER 12 HOURS	AFTER 24 HOURS	AFTER 48 HOURS
C			
LENGTH(mm)			LENGTH(mm)

INITIAL	AFTER 12 HOURS	AFTER 24 HOURS	AFTER 48 HOURS
D			
LENGTH(mm)			LENGTH(mm)

TABLE 3. GEOTROPISM EXPERIMENT.

seedling and orienting it as indicated below. Space the seedlings equally around the beaker.

Seedling B: measure root; orient vertically upward.

Seedling C: measure root; orient horizontally to the right or left.

Seedling D: prior to measuring the root, excise 1 mm at the root tip; orient it horizontally to the right or left.

In Table 3, make a sketch of each seedling to show the appearance and orientation of its root. Record the root lengths.

4. Make observations after twelve, twenty-four, and forty-eight hours. Indicate changes in the length and orientation of the roots with sketches.* At the end of the experiment, measure the main (primary) root accurately, and record the measurement in Table 3.

Explain the changes in each seedling on the basis of auxin distribution in the root.

Demonstrations of Other Tropisms   Examine demonstrations of geotropism and phototropism in stems, and make notes on each demonstration.

THE CHEMICAL CONTROL OF PATTERN IN PLANTS

One of the most challenging problems facing students of both plant and animal development involves the explanation of how pattern (characteristic shape, size, contour, and appearance) comes into being. Leaves are a simple example: Why are they of different shapes in different plants? What selective advantage can there be in maintaining a particular leaf pattern or a particular shape to the plant as a whole? What is the basis for branching pattern, and why does "tapering" tend to be more pronounced in some plants (e.g., conifers) than in others? What controls species-specific size? What causes dwarfness in plants?

We know that plant hormones play an important role in controlling pattern. The branching pattern of roots and shoots, for example, is in part the result of a phenomenon called apical dominance. Lateral buds are sensitive to inhibition by auxin. Since the auxins are produced in active apical meristems, those lateral buds close to such meristems, and thus the branches to which those buds will give rise, are inhibited. The overall plant pattern, of course, depends upon the simultaneous activities of many apical meristems (hundreds or thousands of them in a large tree) and probably on many other factors.

The gibberellins are one class of hormones that seems in some plants to be closely linked to the control of size. First discovered in a fungus that infected abnormally long rice seedlings, gibberellins are now known to occur naturally in plants. They influence many aspects of plant development apart from size: dormancy in seeds and tubers; the initiation of flowering and

---

*Approximate for the first two sets of observations. Do not remove the roots.

Fig. 20-4. Effect of gibberellic acid on cabbage. The plant at left is normal. The one at right was treated with gibberellic acid. *(Courtesy S. H. Wittwer, Michigan State University.)*

fruiting; etc. Though they probably act in conjunction with auxins, they are not active in the growth curvature tests or any of the other standard assays for auxin activity. One type of gibberellin, gibberellic acid, dramatically stimulates stem elongation in dwarf mutants and some other plants that normally undergo little stem elongation (Figure 20-4); such mutants will not respond to auxins. In the following experiment, you will explore the responses of dwarf and normal corn plants to gibberellic acid.

Testing responses to gibberellic acid

1. In the greenhouse or the laboratory, if your instructor so directs, mix equal parts of vermiculite and garden soil sufficient for four 4-inch flower pots. Place this mixture in the four pots to a level about 1 inch below the top. Moisten, but do not soak, the soil mixture in each pot.

2. Obtain 10 germinated "normal" corn seedlings, and place 5 of them gently into depressions in the soil in each of two pots. Cover both sets of seedlings with about a quarter inch of the soil-vermiculite mixture, orienting the green shoot upward through the covering soil layer. Label the pots with your name, date, class, and the designations "Normal—Experimental" or "Normal—Control."

3. Repeat the above procedure with 10 dwarf corn seedlings, using pot labels marked "Dwarf–Experimental" or "Dwarf–Control." Be sure to cover the seedlings with the soil-vermiculite mixture and to orient them as before.

4. Put the pots in two widely separated greenhouse areas designated "Control Plants" and "Experimental Plants." Following your instructor's directions, water the plants thoroughly. Make sure that all information is clearly marked on the pot labels and that the information will not run off when you water the plants.

5. In the following laboratory period, weed out 2 of the seedlings from each pot, leaving the 3 healthiest-looking specimens. Tag and number each seedling, and measure its height in centimeters from the soil to the top of the shoot. Record these data in Table 4 in the column marked "initial length."

6. Using a perfume atomizer or a syringe with a fine needle containing a freshly made water solution of gibberellic acid (100 mg/liter), spray both dwarf and normal *experimental* plants until the leaves have droplets that will almost run off. Avoid wetting the soil with the hormone mixture, and try to apply equal amounts of gibberellic-acid solution to both experimental groups. With a *different* syringe or atomizer, spray the controls with an approximately equal volume of plain water.

7. Measure the height of each plant on each of the four days following treatment and again seven days after treatment. Keep the plants watered, and stake them, if necessary. Record all data in Table 4.

8. At the end of the experiment, analyze your results in whatever way you consider appropriate, comparing normal and dwarf plants in their response to gibberellic acid, and comparing the experimental and control plants in each group.

How soon after application did the response to gibberellic acid occur?

Advance a hypothesis to explain the action of gibberellic acid in correcting dwarfism, and suggest possible means of testing your hypothesis.

CORN VARIETY	TREATMENT	INDIVIDUAL	INITIAL LENGTH (cm)	LENGTH FOLLOWING TREATMENT (cm)				
				DAY 1	DAY 2	DAY 3	DAY 4	DAY 7
NORMAL	G.A. EXPERIMENTAL	1						
		2						
		3						
		AVERAGE						
	WATER CONTROL	1						
		2						
		3						
		AVERAGE						
DWARF	G.A. EXPERIMENTAL	1						
		2						
		3						
		AVERAGE						
	WATER CONTROL	1						
		2						
		3						
		AVERAGE						

TABLE 4. EFFECT OF GIBBERELLIC ACID ON NORMAL AND DWARF CORN SEEDLINGS.

TOPIC **21** ANIMAL DEVELOPMENT

Biological Science, 4th ed.
pp. 788–798.
Elements of Biological Science, 3rd ed.
pp. 477–485.

Development is a continuous process. In sexual organisms, it starts when a sperm fertilizes an egg and ends with death. Between these two extremes occurs a multitude of progressive changes in form and function, ranging from early embryonic development (during which the pattern of the organism is elaborated from a mass of cells) to sexual maturity and aging. Cellular growth and differentiation, metamorphosis (changes in appearance and way of life associated with growth in insects and many other animals), and regeneration (replacement or repair of a lost or an injured part of an organism) are also developmental processes. All organisms develop, and developmental processes are as much the result of adaptation and natural selection as are anatomical, physiological, and behavioral characteristics.

In this laboratory, you will have an opportunity to compare early embryonic development in several living organisms and to observe metamorphosis and regeneration.

Early embryonic development is particularly significant because at this time the pattern of the organism originates. The following summary of the stages of embryonic development is provided to help you follow examples given later in the laboratory.

1. **Fertilization** involves the fusion of nuclei and other events associated with the union of egg and sperm cells (gametes). The product of fertilization is called a **zygote.**

2. **Cleavage** is the term applied to the series of mitotic divisions undergone by the zygote. No growth occurs during early cleavage. The pattern of cleavage—whether the whole zygote or only a part of it divides—varies in different species of organisms according to the amount and distribution of yolk (food material) contained in the egg.

3. **Blastulation** is the formation of a mass of cells that, in most organisms, become arranged into a thin-walled sphere. The central cavity of the blastula is called the **blastocoel.**

4. **Gastrulation** is characterized by morphogenetic movements that lead to the formation of the **archenteron,** or primitive gut, and to the formation of three distinct germ layers of cells—the **ectoderm, mesoderm,** and **endoderm**—from which the organs of the embryo are later elaborated.

5. **Morphogenesis** is the development of characteristic internal and external form. It comes about by the shaping of the germ layers through differential growth, movement, and association of cells in those layers.

We shall compare these processes in four animals: a sea urchin, an echinoderm; a sea star, another echinoderm; a frog; and a chick. Sea-star and sea-urchin embryos have simple and direct blastulation and gastrulation; in frog and chick embryos, these processes are complicated by the presence of a large amount of unevenly distributed yolk.

206

SEA-URCHIN DEVELOPMENT (LIVE MATERIAL)

Sea urchins are scavengers of shallow coastal waters. Observe the adult urchins. Notice the tube feet, each tipped by a small sucker. These enable the urchin to move over the rocks. The spherical body is covered with long spines. Gently poke one of these spines. What happens? What is the advantage of the spines? On the lower surface examine the large mouth with its five teeth. These are used for rasping algae from rocks and tearing the urchin's other food. The sexes are separate; and from five pores on top of the animal, gametes (eggs or sperms) are emitted directly into the ocean, where fertilization occurs. The pores are sometimes difficult to see on a living animal but are readily observable on a cleaned skeleton.

The sexes are not externally distinguishable, but the animals can be sexed and gametes obtained by effecting gamete release in several ways. One easy method involves passing a 10-volt alternating current through the animal. This brings about immediate shedding of gametes, which stops when the flow of current ceases. Another method involves injecting potassium-chloride solution into the body cavity of the animal.

Gametes The instructor will demonstrate how gametes are procured. Make a wet mount of each gamete; if necessary, use Vaseline to prevent evaporation. Be very careful not to mix sperms and eggs or the droppers used to transfer them to slides. (*Note:* It will be necessary to adjust the light carefully in order to see the sperms.)

Fertilization

1. Obtain some fertilized eggs, and make a wet mount. Support the cover slip with Vaseline or with fragments of cover slip.* Successful fertilizations are indicated by the presence of a membrane outside the fertilized egg and a clear space between the membrane and the egg. Unfertilized eggs do not have the space or the membrane.

2. The process of insemination of the eggs occurs too rapidly for you to observe, but you should see many unsuccessful sperms attached to the membrane of the fertilized eggs. If you wish to observe fertilization and the formation of the fertilization membrane, you may do so by mixing a drop of sperm suspension with a drop of egg suspension† on a slide just prior to examination.

Cleavage

The rate of development varies with temperature and with the urchin species; a variety of other environmental factors may also influence it. If your eggs were properly fertilized, you can expect the first cleavage to begin fifty to one hundred minutes after fertilization or later, if the temperature is lower than 20°C. Thirty minutes after fertilization, make fresh wet mounts of fertilized eggs; examine the cytoplasm for evidence of change. Do not leave the slide on the microscope stage unless you turn the light off; heat from a

---

*It is essential to support the cover slip when you observe fertilized eggs since pressure exerted by an unsupported cover slip will prevent normal development.

†Egg and sperm suspensions of proper concentration in natural sea water will be provided by your instructor.

microscope lamp will seriously interfere with development and may kill the embryos before they begin to cleave. With the cover slips properly raised, continue periodically to make wet mounts until you begin to see cleavages.

The cleavages are easily observed in sea-urchin embryos, and with good lighting you may be able to observe some of the internal events that precede the actual cleavages.

The yolk of the echinoderm egg is almost equally distributed throughout; its small amount does not interfere with cytokinesis. It is, therefore, possible for the entire zygote to cleave.

Watch the process of several cleavages carefully, and make sketches. Keep careful records of time.

The table,* below, shows approximate schedules of development for two species of sea urchins. Remember that these are approximations and that the rate of development you observe may be influenced by differences in the room temperature, oxygen supply to the embryos, etc. You may be asked to use batches of previously fertilized eggs to work out a time schedule of development for the species you are using.

STAGE	STRONGYLOCENTROTUS (WEST COAST)	ARBACIA (EAST COAST)
Formation of fertilization membrane	2–5 min.	2–5 min.
First cleavage	70–90 min.	50–70 min.
Second cleavage	100–120 min.	78–107 min.
Third cleavage	130–160 min.	103–145 min.

Blastulation   If developing blastulae are available, make a wet mount and observe them. Notice that they are hollow during this stage. Echinoderm embryos become ciliated, rotate, break out of the fertilization membrane, and begin to swim.

Gastrulation   Gastrulation in the sea-urchin embryo comes about by an infolding or invagination of one side of the blastula called the vegetal pole. The cells of this region gradually work inward and upward until they make contact with the inner side of the cells of the opposite side of the blastula. Thus, the cavity of the blastula, the blastocoel, is almost lost, and the new cavity formed becomes the gut. It is called the archenteron. **Mesoderm** in the sea urchin is formed by outpocketings (evaginations) of the **endoderm** (lining of the archenteron), which eventually break away from the endoderm. The outer layer of the gastrula is called **ectoderm**. All the stages in echinoderm early development are summarized in Figure 21-1, a series of photographs of living sea-star embryos.

*This table is modified in part from D. P. Costello *et al., Methods for Obtaining and Handling Marine Eggs and Embryos* (Woods Hole, Mass.: Marine Biological Laboratory, 1957).

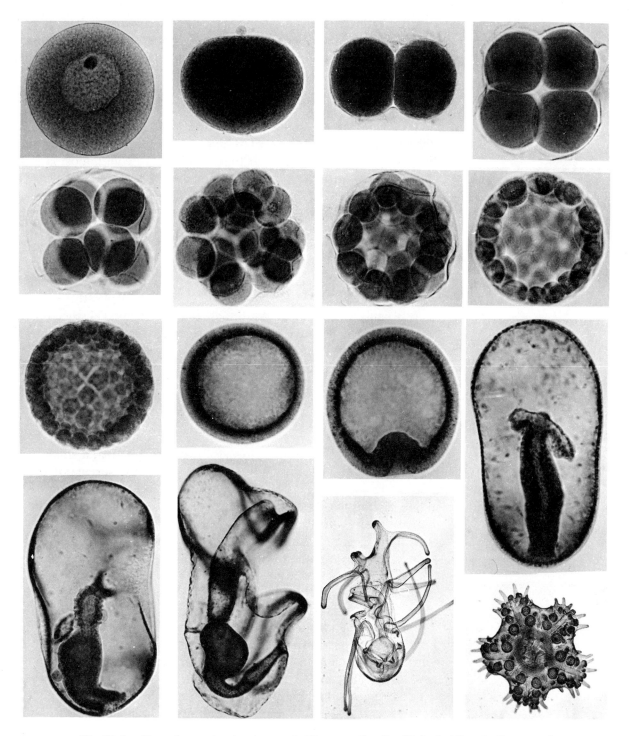

Fig. 21-1. Stages in sea-star development. *(Courtesy Carolina Biological Supply Company.)*

SEA-STAR DEVELOPMENT (PREPARED SLIDES)

Fertilization through gastrulation in the sea star is practically identical to the same stages in the sea urchin. Make sketches as you examine each stage.

Unfertilized Egg   The undivided egg is a nearly spherical cell whose large nucleus and nucleolus are clearly visible. A small amount of **yolk,** or stored food, is present as many small particles distributed throughout the cytoplasm.

Early Cleavage   The two-cell stage, produced by the first mitotic cell division, is easy to identify. The two cells remain attached to each other; their nuclei are usually difficult to see, as is the case with most later stages.

Examine a four-cell embryo. Compare the size of each of its cells with that of the undivided ovum. Do the same for the eight-cell stage. Is there any growth in size? Where does the embryo obtain raw materials for respiration?

Many successive mitotic cleavages produce a many-celled, grapelike cluster called a **morula** (sixteen or more cells). Note that there has still been little increase in the total size of the embryo.

Blastula   As cleavage continues, the embryonic cells become arranged in a sphere, the **blastula.** In sea stars, the blastula walls are usually one cell thick; its central cavity is called the **blastocoel.** (Remember that the specimens on your slides are whole embryos and that you must focus carefully to obtain optical sections showing the internal characteristics of this and later stages.)

Gastrula   Shortly after formation of the blastula, a small depression, or invagination, begins to form at one point on the surface of the embryo. This is the beginning of the critical process called **gastrulation.** If you examine **gastrulae** carefully, you can see that the cells that formed the blastular wall at the end where invagination occurred are larger than those at the other end of the embryo. The difference in size is not very great in sea-star embryos but is more pronounced in many small animals. The smaller cells make up the animal pole of the embryo, and the larger cells make up the vegetal pole; invagination typically occurs at the vegetal pole. As gastrulation proceeds, the invagination becomes larger and larger.

Locate embryos showing the initial steps of gastrulation and other embryos showing many successive degrees of invagination. The latest stages on your slides are embryos in which the inner end of the invagination has begun to expand. (Remember that embryos are randomly oriented on your slides and, therefore, you will view some longitudinally, some diagonally, and some in cross section. Include among your drawings gastrulae shown in both longitudinal and cross sections.)

Gastrulation in the sea star produces an embryo with two primary cell layers: an outer **ectoderm** and an inner **endoderm.** The third primary layer, the **mesoderm,** soon begins to form. Gastrulation, then, eventually results in an embryo with three roughly delineated primary germ layers, an essentially obliterated blastocoel, and a new cavity that is continuous with the outside through the **blastopore** and that will become the lumen of the digestive tract.

Fig. 21-2. Male frog clasping a female frog (amplexus). *(Courtesy Carolina Biological Supply Company.)*

FROG DEVELOPMENT

Eggs are produced in the ovaries of female frogs during the spring and summer months, when food is available and the animals are active. Mature eggs are then held in the ovaries during the winter months and released naturally the following spring. Release occurs during amplexus, contact of the male and female (Figure 21-2), but fertilization and development are external. Frogs brought into the laboratory during the winter or early spring months can be artificially induced to release eggs by injecting them with frog pituitary extract. This procedure apparently simulates the natural hormonal changes that occur in spring. Eggs are released from the ovaries into the coelom and are passed by ciliary action to the oviduct. The eggs may remain in the oviduct for several days, but they can be forced out through the cloaca by gentle pressure on the frog's abdomen.

Active sperms can be obtained by dissecting and macerating the testes of a wintering male frog or by squashing the dissected testes between glass slides. The sperms are then diluted in pond water, checked for activity, and used to fertilize eggs. About an hour before the beginning of the laboratory period, your instructor fertilized some frog eggs. You may obtain them from dishes at the supply desk.

Gametes

1. Obtain a drop of diluted frog sperm from the dish at the supply table. The sperms are very small, and you may be able to see only their sickle-shaped heads. Many of the sperms will be nonmotile, but others will display a spiral forward motion due to the activity of their flagella. (You may occasionally see flagella if the lighting is carefully adjusted.)

2. Obtain some unfertilized eggs from the culture dish at the supply table, using forceps and scissors to cut them away from the rest of the egg mass. Put them into a small dish, and cover them with pond water. Observe them with the dissecting microscope. Notice that the eggs are partially pigmented; the black side contains very little yolk and is called the **animal hemisphere**; the unpigmented, yolky side is called the **vegetal hemisphere**. This unequal distribution of yolk causes some important differences in cleavage and gastrulation between the eggs of the sea urchin and the frog.

Unfertilized eggs are held in random orientation by their sticky jelly coats, and a mass of them will appear mottled because some have the animal hemisphere upward and some have the vegetal hemisphere upward. When it is fertilized, the egg shrinks slightly, and a fertilization membrane very much like a sea urchin's appears. The eggs are then free to rotate within their membranes, and they shift so that the animal pole is uppermost. If you turn your inseminated eggs over when you get them from the supply desk, those that are fertile will rotate in ten to fifteen minutes.

Cleavage

1. The first cleavage will be evidenced by the appearance of a furrow across the zygote's upper surface, the animal pole. Depending upon the water temperature, this will occur 150 to 180 minutes after fertilization. To be certain that you see the furrow, observe your eggs at about five-minute intervals, beginning two hours after fertilization. Once it has begun, the cleavage is rapid.

The crinkling of the egg surface along the cleavage furrow and the disappearance of the furrow after the division is completed are external evidences of mechanical forces involved in animal-cell cytokinesis. Are comparable forces involved in plant-cell cytokinesis?

2. The second cleavage occurs within thirty minutes after the start of the first and will begin at the upper, or animal, pole of the egg. Has the first division been completed when the second begins?

Later embryonic development

Figures 21-3 and 21-4 and your text will help you to understand later development in the frog. Because of the large amount of yolk, gastrulation cannot occur as it does in the sea-urchin embryo, but the end result is the same as in the sea urchin: the three germ layers and archenteron are formed.

The instructor may save your embryos for observation for the following week, or he or she may provide embryos from previous laboratory periods for your observation of later stages in development. If models of frog development are available, be sure to study them sometime during this period.

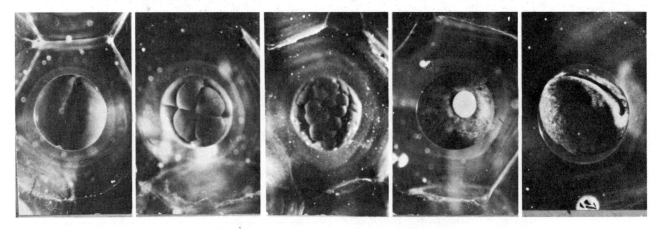

Fig. 21-3. Photographs of frog development; whole embryos. *(Courtesy Carolina Biological Supply Company.)*

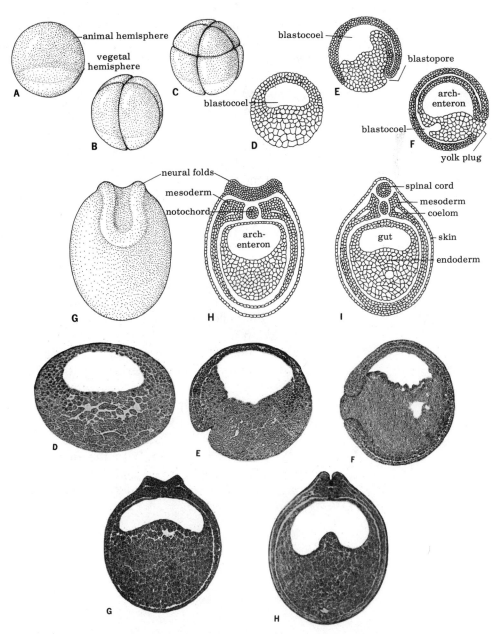

Fig. 21-4. Early embryology of a frog. The upper drawings illustrate some of the stages in early development, sketched whole and in section. The lower photographs show sections through some stages and are lettered to correspond to stages in the drawings. A: Zygote; B, C: early cleavage stages; D: longitudinal section of a blastula; E, F: longitudinal section of two stages in gastrulation; G: an early neurula, showing the neural folds and neural groove; H: cross section of a neurula after formation of the mesoderm; I: cross section of a later embryo, showing the spinal cord and notochord. *(Photographs D, E, F, G: Courtesy Carolina Biological Supply Company; photograph H: Courtesy Ward's Natural Science Establishment, Inc., Rochester, New York.)*

CHICK DEVELOPMENT

In many animals, all embryonic cells are incorporated into the developing new individual. In the development of terrestrial vertebrates, however, a substantial number of embryonic cells develop into temporary structures that make possible embryonic development on land. Reptiles, birds, and mammals have reproductive and developmental adaptations that provide for the metabolic needs, water balance, and general protection of the embryonic individual as it develops within some protective structure (either an egg, as in reptiles and birds, or the uterus of the female parent, as in mammals). You have already studied the mammalian placenta, an adaptation that provides for the growth of the embryo within the uterus of the female parent. Part of the placenta and all of the extraembryonic membranes associated with the embryo, including the amnion, chorion, and allantois and the umbilical cord connecting the embryo to the placenta, are structures composed of cells derived from the fertilized egg. The chick embryo is a convenient one in which to study embryonic membranes and to see some of the important morphogenetic processes in more detail.

Organization of the Egg Examine an unfertilized hen's egg provided by your instructor. The egg cell, or **ovum**, released from the hen's ovary consists of the yellow yolk mass plus a small area of yolk-free cytoplasm called the **blastodisc**, which is located on the surface of the yolk. Because of the large amount of yolk, cleavage occurs only in this disc, and the embryo grows by absorbing raw materials for its metabolism from the yolk (Figure 21-5).

During its passage through the oviduct, several accessory features are added to the ovum. (*Note:* The term "egg" is best reserved for the shelled entity laid by the hen, while the egg cell—that structure released from the ovary—is called an "ovum.") In the first section of the oviduct, a viscous, stringy albumen is secreted and adheres to the ovum. As the ovum moves along, rotation twists the stringy albumen into a pair of strands, which sur-

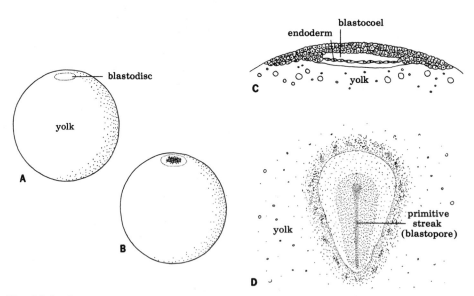

Fig. 21-5. Some stages in the early development of a chick. A: Zygote. A small cytoplasmic disc—the blastodisc—lies on the surface of a massive yolk. B: Early cleavage. There is no cleavage of the yolk. C: Section through a blastula. D: Surface view of a gastrula. Involution of cells along the mid-line of the embryo during gastrulation produces a clearly visible primitive streak, which is essentially a very elongate blastopore.

rounds and projects beyond the ovum. In the next oviduct region, a more watery albumen is added. Finally, two shell membranes and a calcareous shell are applied to the outer albumen.

The egg is a remarkable piece of architecture. The hard, protective shell is porous, permitting gas exchange with the environment. Albumen serves as a water reservoir for the early embryo and as a source of food for the later embryo; it also contains substances that retard the growth of bacteria. The yolk of the ovum contains proteins, carbohydrates, lipids, and vitamins that sustain the metabolic needs of the rapidly growing young chick during its normal twenty-one–day incubation.

Since eggs are held in the terminal portion of the oviduct for variable periods before they are laid, the stage of development at the time of laying varies with each egg. However, cleavage, blastula formation, and gastrulation have usually been accomplished by the time the egg is laid.

Because of the high concentration of yolk in the ovum, cleavage is partial, occurring only in the blastodisc. Gastrulation differs considerably from that in the sea urchin or frog, because the physical forces to which the embryo is subject are very different and because the blastula is a flat plate of cells rather than a hollow ball (Figure 21-5). However, the end result (formation of archenteron and three germ layers) is the same as in the other types of embryos. The yolk of the chick egg remains extraembryonic but becomes enclosed within a membrane, the **yolk sac**, which grows out from the body of the embryo; this contrasts with the frog embryo, in which the yolk is included within certain cleavage cells and eventually becomes a part of the digestive tube.

Opening the egg   Each group of students will be provided with a thirty-three-hour, a forty-eight-hour, a seventy-two-hour, and a ninety-six-hour embryo. Study each embryo with reference to the appropriate descriptive section below and to the appropriate figures—a dissecting microscope is most useful for this purpose. Compare your observations with those of others in your group, so that you see all four stages.

1. To remove your embryo from the shell, tap an egg on the edge of a culture dish about half full of warm Ringer's solution, and force open the halves of the shell under the solution. Do not let the sharp edges of the shell tear or puncture any of the delicate membranes. The yolk mass should shift in the solution so that the embryo is uppermost. Why?

If necessary, add more Ringer's solution to support the embryo properly.

2. After studying the embryo on the yolk, the embryo may be cut off the yolk and transferred to a watch glass for further observation. To remove the embryo from the yolk, place a filter-paper ring over the embryo, pressing the ring down so that it adheres to the membrane. Grasp the ovum membrane outside the embryonic disc with your forceps, and cut the membrane near the edge of the ring. Disturbing the yolk as little as possible, cut completely around the embryonic disc with your finest scissors. Place a watch glass in the culture dish, and gently lift the embryo away from the yolk mass onto the watch glass. Remove the watch glass from the culture dish, and arrange the embryo so that it is free of distortion. Add a little warm saline to the embryo.

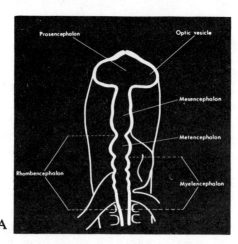
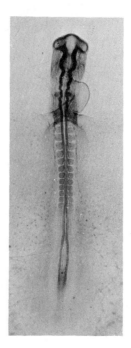

Fig. 21-6. Thirty-three-hour chick embryo. *(A: Courtesy Carolina Biological Supply Company; B: Courtesy Ward's Natural Science Establishment, Inc., Rochester, New York.)*

Thirty-three-hour embryo (Figure 21-6)

At this stage the embryo looks like a light streak in the center of the blastodisc, and you will not be able to see details without removing the embryo from the yolk and examining it with a microscope. Its most outstanding feature is the folding of tissues at the anterior end to form the three major areas of the brain: forebrain, midbrain, and hindbrain. An outline of the foregut should be visible at this time. Anterior to the embryo proper lies a region called the proamnion, so named because it will later join with similar lateral and posterior folds to form the protective amniotic sac. The **neural fold** (forming the hindbrain and spinal cord) forms a groove posteriorly, continuing into the **primitive streak**, the region of the embryo through which cells move into the interior during gastrulation.

The **notochord**, forerunner of the vertebral column, is a supportive rod that can be seen running longitudinally through the center of the embryo. Lateral to the notochord is a series of ten to twelve **somites**, which later develop into such mesodermal derivatives as the skeleton and striated muscles. Anterior to the somites, locate the **vitelline veins**, which will eventually carry food from the yolk to the developing embryo. Parts of these vessels develop into the heart. In order to bring out details, stain the embryo with neutral red found at the supply desk.

Forty-eight-hour embryo (Figure 21-7)

The anterior end of the forty-eight-hour chick embryo has changed considerably from that of the thirty-three-hour embryo. The head has turned so that the right side is up; this torsion of the body will continue until the whole embryo has come to lie on its left side. How far has the torsion proceeded at this stage? Count the number of somites, beginning with the most anterior one. Which is the first somite to show a full dorsal surface to the observer? Note that the anterior part of the head is bent toward the posterior part of

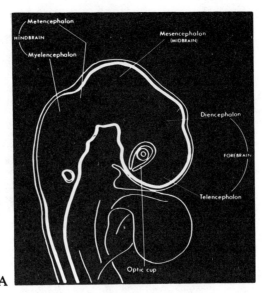

Fig. 21-7. Forty-eight-hour chick embryo. *(A: Courtesy Carolina Biological Supply Company; B: Courtesy Ward's Natural Science Establishment, Inc., Rochester, New York.)*

the head. This bending, or cranial flexure, occurs in the region of the hindbrain. Identify the parts of the brain.

Notice the formation of the heart. Is there any beating or circulation? Have the vitelline veins grown out over the yolk? Has blood formed? Notice, in the pharyngeal area, the development of **gill slits**. What is their significance? Notice the increased folding of the amnion over the head region.

Seventy-two-hour embryo (Figure 21-8)

Torsion and a distinctly C-shaped form are evident. Notice that the embryo is covered by a thin-walled sac, the **amnion**. How far posteriorly does the amnion go? Gently touch the embryo with a proble in order to observe the protection the amnion provides. How many somites have been developed? What changes have taken place in the brain?

Observe the size of the heart and its various chambers. The larger anterior portion is the single atrium, and the ventral posterior portion is the ventricle. Later, the heart will be partitioned into four chambers. Notice the development of the vitelline arteries and veins. Trace the passage of blood through the beating heart.

Notice the development of anterior and posterior limb buds, which will form, respectively, wings and legs. Between the hind limb buds, the **allantois**, which is small at this stage, projects ventrally from the gut. It is one of the four extraembryonic membranes and functions as a depository for excretory wastes; it also distributes blood vessels. The **chorion**, formed at the same time as the amnion, serves with the allantois in the transport and exchange of respiratory gases. Note in Figure 21-9 that the allantois of a bird embryo

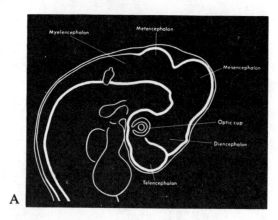

Fig. 21-8. Seventy-two-hour chick embryo. *(A: Courtesy Carolina Biological Supply Company; B: Courtesy Ward's Natural Science Establishment, Inc., Rochester, New York.)*

expands beneath the chorion; the allantois carries blood vessels, a function of which the chorion is incapable. In most mammals, the allantois serves a similar function, expanding beneath the chorion and thus helping to form a highly vascularized double membrane.

Ninety-six-hour embryo (Figure 21-10)

The increased size and vascularity at this stage are obvious. As you opened the egg, you saw blood vessels pressed close against the shell. What is their function? Toward the tail end of the embryo, you should be able to see the fluid-filled allantois. It eventually expands between the embryo and the chorion. As the embryo metabolizes the food material of the yolk, waste products accumulate in this sac. The embryo is now completely surrounded by the amnion.

The liver may be observable as a light-yellow organ filling the anterior

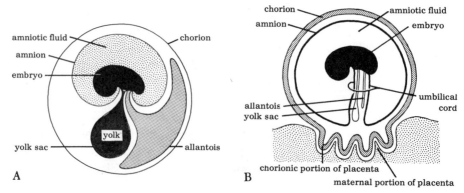

Fig. 21-9. Embryonic membranes. A: The embryonic membranes in a bird's egg; B: a diagram of human embryonic membranes and placenta.

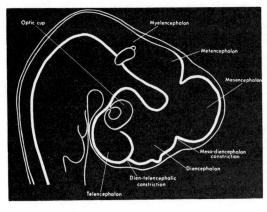
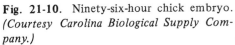
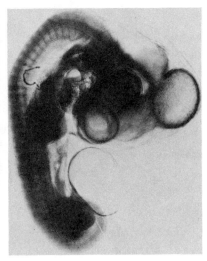

Fig. 21-10. Ninety-six-hour chick embryo. *(Courtesy Carolina Biological Supply Company.)*

portion of the abdominal cavity. Most of the rest of the viscera will appear pinkish-white except for the spleen, a small purple organ.

Notice the increased brain size and differentiation, and the increase in the blood supply to the brain region. The eyes have now become pigmented. Also notice the gill slits and limb buds.

The instructor may be able to provide older embryos for interested students.

EXTRAEMBRYONIC MEMBRANES IN MAMMALS

Your instructor will demonstrate a uterus taken from a pregnant female pig and will discuss some aspects of its anatomy and of pregnancy and birth in mammals. The following outline is provided to help you follow his or her discussion.

Structure:
1. Uterine horns.
2. Body of the uterus.
3. Ovaries.
4. Oviducts, with funnel and ostium at the distal end.

Events in fertilization and early development:
1. Release of eggs from the ovary, and their passage down the oviduct.
2. Release of sperm into the vagina at copulation.
3. Course of sperm up through the uterus into the oviduct, where fertilization occurs.
4. Course of the fertilized eggs down into the uterus, where the eggs eventually implant.

The fertilized egg gives rise not only to the embryo, but also to the three extraembryonic membranes, which may be seen by cutting open a portion of the uterine wall where an embryo is located. The innermost membrane, the amnion, is clear and nonvascularized, and serves as a protective fluid-filled cushion for the embryo. The other two membranes are fused into one: the chorio-allantoic membrane. Note the large number of blood vessels in this membrane, and its foldings which interdigitate with foldings in the uterine wall. The circulation in the embryo and in the chorio-allantoic membrane are connected by the umbilical cord.

The chorio-allantoic membrane and a portion of the uterine wall together form the **placenta**; materials are exchanged between the capillary system in the uterine wall and that in the chorio-allantoic membrane. Note the relatively loose connection between the parts of the placenta in the pig; this connection is much more intimate in the human being and causes some rupture of blood vessels at the time of birth, which would not occur in the pig.

REGENERATION

Regeneration plays an important part in the life of an organism. By this process many important tissues are replaced. The regeneration of lost blood and the healing of broken skin or bone are examples. Some species are capable of regenerating entire organs or regenerating an entire organism from a part.

Regeneration is very similar to embryonic development. Unspecialized cells rapidly divide, each eventually becoming some particular part of the organism. Regeneration entails the same problem as does embryonic morphogenesis, where unspecialized cells differentiate. During both embryonic development and regeneration, a strong polarity exists where the anterior region develops at a greater rate than the posterior region. In the following study you will have an opportunity to compare rates of development in different regenerating regions.

You will work with fresh-water planarian *Dugesia.* We have selected these animals for the study because they have a high potential for regeneration. Planarians have been known to regenerate a whole new organism from a middle section. As a rule, the more anterior the origin of the section, the more likely it is to regenerate entirely. Why?

You will be provided with a planarian, vial(s), and a separate container of pond water. During the next two weeks, you will watch this planarian regenerate at cuts you design.

Cutting the worm
1. Obtain a planarian from the stock container at the supply desk.

2. Carefully place the planarian on a moistened cork or on a piece of glass. When the worm extends itself on the flat surface, make the necessary cuts with a razor blade. It is best to perform all operations while observing the worm with a dissecting microscope. Suggested cuts are outlined in Figure 21-11.

3. Transfer the cut sections into the vial(s) provided. Fill the vial(s) with pond water, place the cork or cap on tightly, and closely observe the planarian every day.

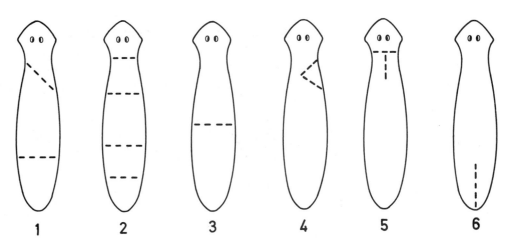

Fig. 21-11. Suggested cuts on a planarian.

Care of the worm

1. During the next two weeks of close observation, do not feed the planarian. Planarians that have been operated on are hardly in a position to begin feeding right away; giving them food will only foul the water, leading to death and disintegration of the worms or fragments.

2. Change the water daily. Use only clean spring or pond water provided by your instructor. Keep the planarian in a cool place, protected from strong light.

3. Any cuts intended to separate two or more portions without detaching them completely from the rest of the worm will have to be renewed every few hours for at least half a day, to prevent their re-fusing.

4. Keep a record of your observations. Unless the cuts made in the operations are recorded in sketches, you are likely to lose track of the origin of some pieces by the time they have undergone at least some regeneration.

5. Compare rates of regeneration between different sections, studying other students' data where necessary. Is there a gradient between the anterior and posterior regions?

Look up the phylum and class of each of the animals you have studied in this period, and, if they are not already included, enter each of them in the appropriate space in Appendix 2.

QUESTIONS FOR FURTHER THOUGHT

1. Compare early cleavage in a sea urchin, a frog, and a chick.
2. Since the parents of sea urchins, sea stars, and many other marine animals never come into contact, gamete release in adjacent animals must

be reasonably simultaneous. What factors do you suppose trigger the simultaneous release of gametes under natural conditions?

3. Relative to their adult size, birds' eggs are very large when compared to those of echinoderms. What problems of survival necessitate the large size of a bird's egg, and how are these problems solved by echinoderms?

4. Why is rate of development a function of temperature?

5. From what two structures is the mammalian placenta derived?

6. What is a germ layer? What structures are formed from each germ layer?

7. In some fly larvae, growth occurs by cell enlargement rather than by cell division. What adaptive value would there be to this arrangement?

TOPIC **22** EVOLUTION

Biological Science, 4th ed.
Chap. 34, pp. 875–877.
Elements of Biological Science, 3rd ed.
Chap. 26, p. 600.

Throughout this course we have attempted to interpret all topics in the light of modern evolutionary theory. This has given us a framework within which we can better understand factual detail and perceive its significance. We have emphasized that the theory of evolution, one of the unifying principles of biology, involves two basic ideas: (1) that the characteristics of populations of organisms change over periods of time, and (2) that the direction and rate of evolutionary change is determined by natural selection. You have studied the genetic basis for evolutionary change and should now understand how natural selection brings about shifts in the composition of the gene pool of the population. This laboratory should clarify further some of the concepts in evolutionary theory.

In this laboratory you will examine a series of demonstrations illustrating evolutionary principles. The exhibits are arranged in the order of their treatment in this Topic. Study each carefully, and answer the questions asked. Sometimes there is no absolute answer for a question, and you are then expected to be able to defend whatever answer you select. The purpose of this exercise is not to have you memorize specific facts about particular animals and plants exhibited. Rather it is to help clarify some basic concepts of evolutionary theory by giving you personal contact with actual examples of evolution. The organisms in these exhibits were chosen because they provide good illustrations of important ideas and because they are easily obtained; many other organisms might just as well have been used. You should, therefore, concentrate more on the general principles illustrated than on the names and facts involved.

Some of the exhibits listed below may not be available to you, and others not listed here may have been added. Follow your instructor's directions concerning the latter exhibits.

PALEONTOLOGICAL EVIDENCE FOR EVOLUTION

Fossils provide scientists with some of the best evidence about the past history of life on earth, and at times the fossils make it possible to trace in detail the evolutionary changes that have taken place in various lines of organisms. Some of the major trends in organic evolution should have become apparent to you as you used Appendix 2 and will be further clarified later in the course when we take up the different groups of organisms. These major trends did not become apparent to biologists all at once but have slowly emerged from the painstaking integration of hundreds of different bits of evidence accumulated through the years by a multitude of individual research efforts, each dealing with the evolution of a specific group of organisms.

The evolutionary history of the horse family is particularly well known because its fossil record is more complete than most.

1. Examine the large mount illustrating horse evolution. This display shows the characteristics of the skull, forelegs, hind legs, and teeth of four horses; each is shown in its proper geological period (i.e., Eocene, Oligocene, Miocene, etc.). Between the lower Eocene *Eohippus* and the modern *Equus* you can recognize many fossil stages. The chief changes you should note are:

    a. Increase in overall body size.

    b. Lengthening of the distal portion of the legs.

    c. Reduction of lateral digits.

    d. Increase in length of the anterior part of the skull.

    e. Increase in depth of the molars, and increase of their grinding lophs (ridges).

    f. Increase in resemblance of the premolars to the molars.

    g. Completion of the postorbital bar of skull (i.e., the bar behind the eye socket).

Doubtless, change has also taken place in many other characteristics; we know, for instance, that there has been a progressive increase in the size of the neocortex (neopallium). This was not determined directly, since soft tissues do not become fossilized readily; but cranial casts can be used to calculate such a character.

2. Examine the series of fossils of other organisms on display, and read the explanatory cards beside them.

DIVERGENT EVOLUTION

In studying the evolution of horses, you focused attention on a few representatives, each from a different geological epoch. Stacked one next to the other, they may give the misleading impression that the evolution of the horse family progressed in a single direct line, starting with *Eohippus* and terminating with *Equus*. This oversimplification, if not properly qualified, amounts to a falsification. In reality, things happened in a far more complex, yet more meaningful, way. At any one time there lived not just one type of horse, but a diversity of them. Some types were destined to become extinct because they failed to evolve satisfactory adaptations, whereas others, which did adapt, were able to survive by successfully meeting new environmental requirements. The representatives we chose for study belonged to specific strains that are known to have survived through change. In other words, they did not belong to evolutionary "blind alleys." That is why when we consider them alone, disregarding all others, we are left with the impression of a straight, direct ascendancy. It is important to note that only a fraction of the total number of species of any one group of organisms living at a certain time is destined for survival. We know this has been true in the past because we have ample substantiating evidence from fossils; and there is certainly every reason to believe that it will also be true for species living today.

At any one time, then, most groups show considerable diversity. A series of demonstrations has been set up to help you understand such adaptive variation, which results from divergent evolution—i.e., increasing dissimilarity between two or more populations that were initially one. (What is the difference between divergent evolution and parallel evolution?)

Since we shall consider diversity at many levels of the taxonomic hierarchy, you should be familiar with the chief categories of this hierarchy. These are:

Kingdom
Phylum or division
Class
Order
Family
Genus
Species

This system of classification provides a method whereby biologists pigeonhole and extract meaningful pattern from the innumerable diverse organisms with which they must work. In essence, it is simply an outline, much like one you might use to help you through your history or chemisty course. Such classification could be based on almost any characteristic of the organisms being studied, but scientists have chosen to use common ancestry as the classification principle. Can you name four or five major sources of information a biologist might use in determining ancestry?

Divergence among phyla of a single kingdom

Examine the demonstration showing members of five different animal phyla. Presumably all animals share a common, remote ancestry, but they have diverged greatly and can now be divided into a number of large categories (phyla) based on degree of relationship. The gross characters used for separation at this high level of the taxonomic hierarchy should be apparent from even a superficial examination of the five specimens shown here.

1. Would you expect that an earthworm belongs to the same phylum as specimen A, B, C, D, or E? On what characters do you base your decision?
2. Would you expect that a crayfish belongs to the same phylum as specimen A, B, C, D, or E? Again, give reasons for your decision.

Divergence within a phylum

Examine the demonstration of the classes within the phylum Mollusca. Most members of this phylum have shells, though the forms of the shells vary greatly. Most cephalopod molluscs, however, have no external shell, though their ancestors, known now only from fossils, were shelled.

1. Contrast the shells of class A (Amphineura) with those of the other classes.
2. Contrast the shells of class B (Gastropoda) with those of the other classes.
3. Contrast the shells of class C (Pelecypoda) with those of the other classes.
4. Contrast the shells of class D (Scaphopoda) with those of the other classes.

5. What adaptive advantages might the members of class E (Cephalopoda) have gained by losing their shells?

Within some of these classes, more diversity can be seen on demonstration. This is divergent evolution within classes instead of among classes.

Divergence within a Class Before proceeding any further, we should clarify the meanings of a few terms commonly used in evolutionary discussions. Two structures (or two characteristics of any type, for that matter) are said to be **homologous** if they evolved from a common ancestral structure. If two characteristics are functionally similar but evolved from different ancestral characteristics, they are said to be **analogous**. It is necessary to specify in what sense either of these two terms is used in any given instance. For example, the flippers of whales and seals evolved from the front legs of land-mammal ancestors. Thus, they are homologous in the sense that both are forelimbs, with the same basic bone structure as other vertebrate forelimbs; but the modifications that make them flippers are not homologous, and they must be said to be analogous with regard to their flipper characteristics.

Two other terms, **specialized** and **generalized**, also need clarification. These terms are used to indicate the relationships of organisms, or particular characteristics of organisms, to their environment. "Specialized" means adapted to a special, usually narrow, way of life; "generalized" means broadly adapted to a greater variety of different types of environments and ways of life. For example, the rear legs of a frog are specialized for a particular type of locomotion on land; the rear legs of a dog, while specialized in many ways themselves, can be regarded as more generalized than those of a frog.

The terms **primitive** and **advanced** are used in comparing organisms, or characteristics of organisms, to the ancestral type. "Primitive" means first, original, like the earliest member; "advanced" means the opposite. A word of caution: "primitive" and "advanced" do not imply any value judgment that one is "better" than the other, nor do they imply that one is more complex than the other; they refer *only* to relative sequence in time.

Divergence within various classes

1. When studying the molluscs, you saw that main evolutionary lines within a phylum can diverge so much that they come to be considered as separate classes, and you saw also that there is divergence within each class. Now examine representatives of the vertebrate class Reptilia. All reptiles are related at the class level, yet they show striking differences that provide a good example of strongly divergent evolution. The example would be even more impressive if we could include the many diverse dinosaur types that once roamed the earth, all of which were reptiles.
   a. The fossil record indicates that the ancestors of the reptiles were four-legged amphibians without shells (their appearance resembled that of modern salamanders). Would you say that the legs of turtles and lizards are homologous structures? Explain your answer.
   b. Which term, "homologous" or "analogous," best describes the relationship of the shell of a turtle to that of a snail?
   c. Which of the reptiles on demonstration would you consider the most primitive in overall body plan? Why?

d. Which of the reptiles would you consider the most advanced in overall body plan? Defend your answer.

e. Remembering that reptiles evolved from amphibians, which would you say is more primitive—for a reptile to have legs or to lack them?

f. Which is more specialized—for a reptile to have legs or to lack them?

2. Another striking example of divergence within a class is seen in fish. On display are representatives of unusual types of fish, which should give you some idea of the great diversity within this class.

Examine each of the fish, and briefly mention ways in which it differs from the perch (on demonstration), an example of the more usual fish form.

3. Examine the demonstration of the forelimbs of various members of the class Mammalia, noting both similarities and differences. (Three forelimbs—bird, frog, and turtle—included in the display are not mammalian.) Identify each bone in the human arm, remembering your study of the skeleton in Topic 13, and then try to locate the same bones in the limbs on display. You will see that bones differ greatly in length, thickness, degree of fusion with each other, etc., in the different animals. Having compared the bones of the different species (it may help you to make a table showing your findings), try to explain your data on the basis of the way of life of the animals involved. The object here is not to memorize the bones of the specimens examined, but to understand the way the same basic structures can be molded by evolution into a great variety of types.

Divergence within an order

1. Examine the display skulls of various members of the mammalian order Carnivora. All members of this order possess certain skull characteristics in common, particularly the development of the canine teeth. This display includes representatives of four carnivore families: dog, cat, bear, and mink. Some of the differences among these should be apparent to you.

a. Compare the four families with regard to general shape of skull, tooth characteristics, etc.

b. What animals (not shown here) would you think are in the "dog" family?

c. What animals (not shown here) would you think are in the "cat" family?

2. The rodents comprise another order of mammals. Examine several rodent skulls.

a. Compare these with the skulls of carnivores, paying particular attention to the teeth.

b. What do these differences indicate about rodents' diets?

3. A third mammalian order, Artiodactyla, includes all hoofed animals with an even number of toes: cow, deer, sheep, goat, giraffe, pig. Examine the skulls on display, and note similarities and differences.

a. Contrast the teeth of members of this order with those of carnivores.

b. What can you tell about the diet of members of this order by examining their teeth?

4. A fourth mammalian order, Primata, includes lemurs, monkeys, apes,

and men. Examine skull casts of a gibbon, an orangutan, a gorilla, and a chimpanzee, and compare these to the human skull.

   a. Characterize the most striking differences between skulls of this order and those of carnivores and rodents (note the teeth particularly).

   b. Which of the other primate skulls on display do you think most closely resembles the human skull? Why?

   c. What major changes would be necessary to convert one of these skulls into one like that of man?

5. The insect order of beetles, Coleoptera, provides another good example of adaptive radiation within a single order. Examine the case containing many different beetles. The block in the lower right-hand corner contains other insects, not members of the Coleoptera, which you may use for comparison in answering some of the questions below.

   a. What characteristics do all the beetles have in common that distinguish them from the other insects shown?

   b. Each block of beetles contains members of a single family. How do the first and second families differ from each other?

   c. How do the third and fourth families differ from each other?

   d. Examine the six specimens on the pinning block outside the case. All of these are members of the class Insecta, but not all are beetles. Judging from your generalizations in 5a, which are Coleoptera?

Divergence within a family

You should have some idea of families as a result of your examination of carnivores. Members of the cactus family provide a good example of divergence within a family. All clearly resemble each other more than they resemble dandelions, oak trees, or grasses; but in spite of their basic similarities, the differences among the species on display are obvious.

   1. Several members of a family of butterflies are on display. Both their similarities and differences should be obvious.

   2. An example of a family with diverse members is the one that includes pineapple, Spanish moss (not really a moss), and many epiphytic plants (aerial plants growing on other plants but not parasitic on them—very common in the tropics). Examine members of this family.

Divergence within a species

Examine the display showing twenty-four different shells of the bivalve mollusc *Donax variabilis,* the cochina clam. Note the striking variations in both color and pattern. Not only are all these shells of one species, but all were collected at one time and at a single locality. This, then, is an example of intraspecific variation that is not correlated with geography.

Sexual dimorphism

One type of intraspecific variation that should be distinguished from the types already seen is sexual dimorphism—i.e., the male and female of a single species look different. Examine the two sexes of the birds or insects on display. What might be the adaptive significance of the differences you see?

CONVERGENT EVOLUTION

A constant problem facing biologists is determining when similarity of appearance indicates actual phylogenetic relationships and when it is simply a reflection of similar selection pressures acting on unrelated organisms. When the latter is true, we term it "convergent evolution." A series of demonstrations has been set up to help you understand convergence.

Demonstrations of convergence

1. Examine wings of bat and wings of bird. The two vertebrates evolved wings independently, though in each case the forelimb was the chief structure modified. Compare these wings with those of an insect, an animal in a different phylum from the other two. The insects' wings did not evolve from legs, and are very unlike the wings of birds and bats in basic structure.
   a. Should you use the term "homologous" or "analogous" when comparing the wings of birds and insects?
   b. In what sense are the wings of birds and bats homologous?
   c. In what sense are the wings of birds and bats not homologous?

2. Jointed legs have arisen convergently twice in the animal kingdom: in the phyla Arthropoda and Chordata. Note that such legs would be functionally efficient only in animals with hard skeletal systems.
   a. Are the legs of grasshoppers and squirrels homologous or analogous?
   b. Use the appropriate term—"homologous" or "analogous"—with reference to the legs of lobster and man.
   c. Use the appropriate term—"homologous" or "analogous"—with reference to the legs of raccoon and man.

3. Many unrelated animals have convergently evolved "wormlike" bodies. Examine a series of such animals, including earthworm, various insect larvae, millipeds, centipeds, snake, salamander, etc.

   For what sorts of habitats is this body shape particularly well adapted?

4. Mimicry is an especially interesting type of convergent evolution. Here selection causes unrelated species to evolve similar appearances involving protection from predators. The species that are naturally protected by virtue of some "unpleasant" characteristics such as taste, smell, sting, etc. are called the model species; they often are brightly colored or conspicuous in some other way, thereby making it easier for predators who have experienced their unpleasant features to recognize them and avoid them in the future. Species that are not naturally protected by some unpleasant characteristic of their own often mimic the appearance of model species and consequently avoid predation because the predators can't distinguish them from the unpleasant models. This phenomenon is fairly common among insects, as the display mounts indicate. The displays show the model species on the left with the mimic species to their right. Often only one sex of a given species is a mimic, the other sex appearing entirely different (providing more examples of sexual dimorphism).

   There are two chief types of mimicry, Batesian and Müllerian. In the first type, only the model species is distasteful (i.e., tastes bad or smells bad or stings, etc.), and the mimic species, which has no such defensive mechanism, gains protection from predators by being mistaken for the model.

In Müllerian mimicry, all of the species involved are, essentially, both models and mimics—that is, all have some defensive mechanism. But if each had its own characteristic appearance, the predators would have to learn to avoid each species separately, thereby making the learning process a demanding one and also necessitating the sacrifice of some individuals of each species to the learning process. If, however, species convergently evolve more and more toward one appearance type, they come to constitute a single prey group from the standpoint of the predators, and avoidance is more easily learned.

5. A well-known example of convergent evolution between two different plant families is that involving cacti and euphorbias. The two types of plants look so much alike that the average person normally calls both "cacti," yet they are really not closely related. Can you think of reasons why these two plants, which live in the same sort of habitat, might have evolved the particular characteristics that both possess in common?

Biological Science, 4th ed.
pp. 1029–1060.
Elements of Biological Science, 3rd ed.
pp. 621–641.

One major aim of this laboratory—and, indeed, of your entire laboratory experience—is to help you understand the lives of other organisms. Learning the names and functions of parts of animals may be of some interest, but it is not really meaningful until you understand the significance of each anatomical, functional, and behavioral adaptation that helps fit an organism to life in its particular habitat. Occasionally we will pose questions to guide your thought, but they should be considered only as guides; most of your questions should arise from your own study of the materials provided. Sometimes the answers will just be common sense, and sometimes they will be generated by the results of your experiments. In either case, training yourself to be a questioner will be rewarded by a much deeper and more meaningful comprehension of organisms than could ever be the case through rote memorization.

Earlier in this course, as we reviewed the general problems facing all living organisms, our attention was often directed to the remarkable adaptations with which the vascular plants have met the challenges of terrestrial existence. While we marvel at these adaptations and recognize the importance of vascular plants in terrestrial communities as shelter and food for numerous other organisms, we should not allow ourselves to lose sight of the fact that vascular plants are only a small representation of the total number of living organisms.

In this Topic, we will examine the Monera and Protista. The Monera are structurally simple, procaryotic organisms. Procaryotic cells lack a membrane around the nucleus, have no mitochondria, endoplasmic reticulum, Golgi apparatus, or lysosomes. Monerans reproduce by binary fission rather than by mitosis or meiosis. The kingdom Monera includes the Archaebacteria, the "ancient" bacteria, and Eubacteria, the much more common "true" bacteria. These are the oldest, as well as the hardiest, of all living organisms. They do not simply tolerate some extreme environmental conditions, but thrive in these conditions.

The kingdom Protista is composed of three groups: animal-like, plant-like, and funguslike organisms. The members of this kingdom are primarily unicellular or colonial eucaryotic organisms (or similar multicellular or multinucleate ones). Protista exhibit tremendous variation in life cycles and cell organization but are usually confined to an aquatic or a very moist habitat.

MONERA

Bacteria Bacteria play an extremely important role in our lives; they are crucial in medicine, agriculture, and diseases, as well as in a number of industrial processes. You will examine prepared slides of three different types of bacteria and then will prepare your own slide from cultures provided by your instructor. Bacterial cells are described as rod-shaped, spherical, or helically coiled. The rod-shaped bacteria are called **bacilli**, the spherical bacteria **cocci** (singular, "coccus"), and the helically coiled **spirilla**.

Bacilli

Obtain a slide of *Clostridium botulinum.* These rod-shaped bacteria are fermenting bacteria. These organisms are **obligate anaerobes**—that is, they cannot live in the presence of oxygen. However, when conditions are unfavorable for growth, these bacteria are able to form spores and germinate again when conditions are favorable. If food is ingested on which *Clostridium botulinum* has grown, the result is botulism, or food poisoning, produced from the powerful toxin released from these cells. This toxin blocks the release of acetylcholine at the neuromuscular junction in humans.

With high power and the oil-immersion lens, observe this specimen and those described below. Make drawings, labeled appropriately, of each of the slides. Note whether the cells are found individually, in chains, or in colonies.

Coccus

Streptococcus lactis, a nonpathogenic bacterium, is found in milk products, especially cheese, and is responsible for the normal souring of milk.

Spirillum

Rhodospirillum rubrum, found in stagnant waters and mud, is an anaerobic photosynthetic bacterium that the process of photosynthesis is thought to have originated billions of years ago. In bacteria, the photosynthetic pigments and enzymes are not in chloroplasts but in chromatophores (organelles composed of vesicles), and the organism only possesses Photosystem I.

Gram staining

When studying bacteria, you should have some idea of the morphology of an organism—for instance, the shape of the cells, their size, and any distinguishing characteristics. For this purpose, simple staining techniques provide one of the most useful tools. Although the stain allows you to see the bacteria, it does not permit you to distinguish between morphologically similar organisms.

Certain groups of bacteria may be distinguished by the Gram stain, which actually involves two stains: crystal violet and safranin. Almost all bacteria stain can be stained with crystal violet, which results in a purple color. Those bacteria that retain the violet color are classified as Gram-positive; those that decolorize but take on the counterstain (safranin) are called Gram-negative. The difference in staining is due to difference in composition of the cell wall. Most of the bacteria that affect the health of humans—such as the bacteria that spoil food, cause pneumonia, or sour milk—are Gram-positive.

Using sterile techniques, as demonstrated by your instructor, prepare slides for staining in the following manner:

1. Place a drop of distilled water on a clean slide.
2. With a sterile inoculating loop, remove a small amount of material from a colony, and stir the loopful of cells into the drop of water. Spread the drop to cover an area of about 1 square centimeter, and let the slide air dry. This dry bacterial film is referred to as a "smear."
3. "Fix" the smear by passing the slide three times close to a low flame, but keeping the smear away from the flame.

4. Cover the smear with the crystal-violet stain, and let the smear stain for one minute.

5. Rinse the stain with tap water.

6. Cover the smear with iodine solution, and let the slide sit for one minute.

7. Wash the iodine solution off with tap water, and gently blot the slide dry.

8. Decolorize your smear with 95% alcohol. Let the alcohol run down the slide until no more stain washes out.

9. Stain the smear with safranin for twenty to thirty seconds, rinse the slide with tap water, and blot the slide dry. Examine your slide under the oil-immersion lens.

Make a sketch of the bacteria, and record whether they are Gram-negative or Gram-positive.

Cyanobacteria   These organisms, previously called blue-green algae, are aerobic photosynthetic bacteria. Like in the anaerobic photosynthetic bacteria, the pigments for photosynthesis (the same ones as are found in the higher plants) are located in thylakoids, flattened vesicles similar to the chromatophores. These organisms also contain the pigments **phycocyanin** or sometimes **phycoerythrin**.

Anabaena   Obtain some *Anabaena* (Figure 23-1), and make a wet mount for microscopic observation. Note that there are three types of cells in the filament: **vegetative cells; heterocysts**, which appear slightly larger and denser than vegetative cells; and **akinetes**, the largest of the cell types. When conditions are unfavorable, the vegetative cells develop into thick-walled spores called akinetes. The heterocysts are specialized for nitrogen fixation. If *Anabaena* is growing in an environment where the level of nitrogen-containing substances is limiting, then heterocysts begin to develop. The enzyme involved in nitro-

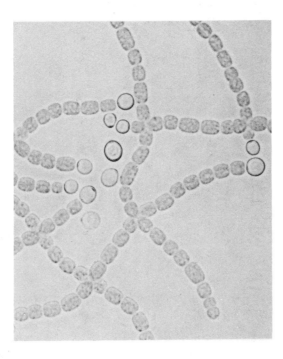

Fig. 23-1. *Anabaena* with numerous heterocysts—structures specialized for nitrogen fixation.

gen fixation can only function in an anaerobic environment, and these cells (heterocysts) are specially adapted to maintain this type of environment by having no Photosystem II.

Many of the Cyanobacteria are bound together in a gelatinous sheath. Is *Anabaena* enclosed in a sheath? What advantages might there be in this type of organization (in contrast to individual free-living cells)?

PROTISTA

Animal-like Protista

Paramecium

Most of the Protozoa and a variety of cells in multicellular organisms—for example, white blood cells—feed by engulfing smaller organisms directly into temporary vacuoles. These food vacuoles circulate in a characteristic manner within the cells, and enzymes are secreted into them for digestion. Some Protozoa, such as *Amoeba,* can form vacuoles anywhere on the cell surface; others, like *Paramecium,* have regions of the cell specialized for vacuole formation.

Make a wet mount of *Paramecium* culture into which you have stirred a drop of methyl cellulose and a tiny bit of Congo-red yeast mixture (just enough to color the mixture pale pink—toothpicks are available for mixing). Examine the slide under low and high powers, with reference to Figure 23-2.

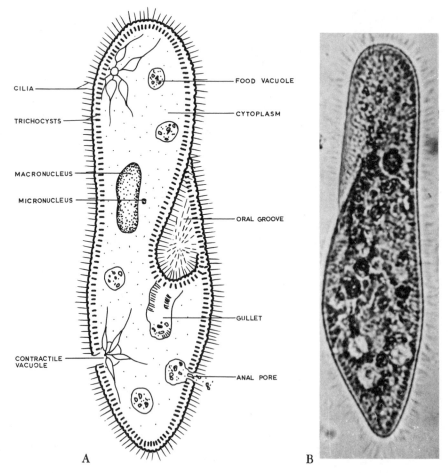

Fig. 23-2. *Paramecium. (Courtesy Carolina Biological Supply Company.)*

Locate the **cilia,** fine cytoplasmic threads projecting from the surface of the animal. Very long cilia sweep minute particles (stained yeast, in this case) along an **oral groove** and into a **gullet,** at the end of which food vacuoles are formed. Locate a more or less stationary animal, and watch the collection of yeasts at the end of the gullet, formation of a food vacuole, and the manner in which the vacuole circulates inside the cell. If you are a patient observer and the animal you choose is quiet enough, you may be able to map the route taken by the vacuole. Very observant students will find that the color of the vacuole changes on its route through the cytoplasm. Your instructor will have more information on this reaction between the dye on the yeasts and the materials within the food vacuole secreted by the *Paramecium.* (What sorts of substances would these be?)

Vacuoles of this kind are formed anew each time food is ingested; these animals lack permanent structures to digest food. Thus, food vacuoles are the functional counterparts of the digestive systems of multicellular animals.

Is this intracellular or extracellular digestion?

Do you think *Paramecium* and *Amoeba,* living in the same habitat, ever compete directly for food?

Stentor

Stentor, a relative of *Paramecium,* has organelles for feeding, digestion, locomotion, water balance, oxidative metabolism, and integration; in addition, it exhibits a remarkably complex behavior. These same activities in a human being would require the cooperative action of organ systems composed of many millions of specialized cells. In this unit, you will review some of the complex internal structure that characterizes Protozoa, and you will explore several aspects of *Stentor's* behavior toward food.

Obtain several organisms on either a depression slide without a cover slip, or on a plain slide with a cover slip raised by Vaseline or bits of glass. Study the structure of *Stentor,* and note the following with the aid of Figures 23-3 and 23-4:

1. Trumpet-shaped body.
2. Frontal field (the region where food is ingested).

Fig. 23-3. Living *Stentor.* This photomicrograph was made with oblique light. *(Courtesy Carolina Biological Supply Company.)*

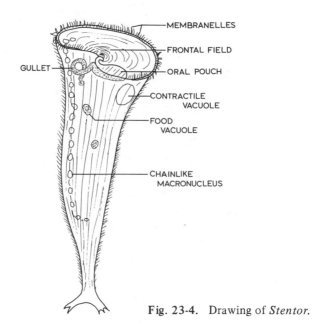

Fig. 23-4. Drawing of *Stentor*.

3. Band of membranelles (fused cilia).
4. Oral pouch.
5. Gullet.
6. Food vacuoles.
7. Contractile vacuoles.
8. Chainlike macronucleus.

Use the dissecting microscope and the low power of your compound microscope; experiment with different types of lighting until you find a set of conditions that gives you clear visibility of the organism. Later, you will need to be able to observe the feeding current created by the beating of membranelles and cilia around the frontal field. Find animals that are attached; if the animal you are observing swims away, locate another one that is attached.

Feeding current   Lift one edge of the cover slip with a dissecting needle, and, on the end of a toothpick, introduce a very small amount of yeast suspension stained with carmine or with Congo red. Carefully watch the current created by the beating of the membranellar band. This current serves to bring suspended particles into the gullet and may be reversed or stopped under certain conditions. Watch the current for some time, until you can describe its shape and direction. Using arrows and lines, add this information to Figure 23-4.

Food selection   By carefully observing animals on the slide, you may be able to detect evidence of discrimination between dye particles and yeasts. If you can see discrimination, where does the sorting occur? Is the selection complete, or are some indigestible dye particles taken into the animal?

A better demonstration of food selection will occur if unfed animals on a new slide are offered a combination of *Euglena* (a green flagellate) and indigestible carmine-dye particles. By observing the formation of food vacuoles, it should be simple to determine whether selection is occurring, whether any indigestible particles are taken in, and, if so, the approximate ratio between digestible and indigestible materials.

Why would food selection be adaptive for the animal? What "criteria" might it use in exercising selection?

Behavioral responses to nonnutritive material

It is possible that an animal like *Stentor* might, on occasion, find itself in a situation where the beating of the membranelles brings in only nonnutritive material. In such circumstances, a behavior pattern enabling detection and escape would be advantageous. With a clean toothpick, add some of the carmine suspension to a fresh mount of unfed *Stentor*. Carefully watch and record any changes in the feeding current and in the posture and movements of the animal. Describe the reaction sequence. Compare the reactions of your animal with reactions observed by others in the class. Are the responses consistent? How would each one be adaptive?

Other Protozoa

The Protozoa are customarily divided into five phyla:

1. Mastigophora.
2. Sarcodina.
3. Ciliata.
4. Sporozoa.
5. Cnidospora.

Review and list the characteristics of each phylum. Your instructor may provide demonstrations of protozoans in addition to those you have already seen. Determine the phylum of each of these animals, and enter it in the appropriate space in Appendix 2.

Plantlike Protista

Diatoms

The most conspicuous feature of this group of Protista is the exquisite patterns of their shells or valves (Figure 23-5). These patterns are formed from the pores in the shell which provide contact between the protoplast and the external environment. The silica-impregnated shells fit together like a box; when a cell divides by mitosis, each daughter cell retains half of the parent shell and constructs the other half. Diatoms are similar to higher organisms in that they spend most of their time as diploid organisms and go through meiosis just before gametogenesis.

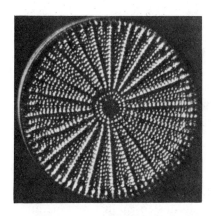 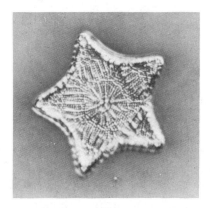 

Fig. 23-5. Three species of diatom. Left: *Arachnoidiscus ehrenbergi;* middle: *Triceratium pentacrinus;* right: *Trinacria regina. (Courtesy Turtox/Cambosco, Macmillan Science Co., Inc.)*

Prepare a wet mount of a drop of the living diatom culture, or examine prepared slides. Diagram several of the organisms.

Funguslike Protista

Physarum   Slime molds are found living in woods on decaying vegetation. They exhibit both funguslike and protistlike characteristics during their life cycle (Figure 23-6).

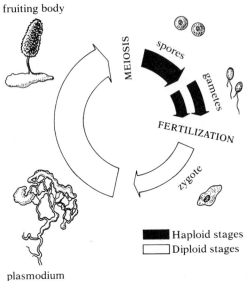

fruiting body

plasmodium

Fig. 23-6. Life cycle of a true slime mold.

Physarum is classified as a **true slime mold** (division Myxomycota). You will observe the **plasmodium** stage, which is a multinucleate amoeboid mass, and the **fruiting body**, the funguslike stage. The fruiting body develops when conditions are no longer favorable for plasmodium growth. Examine the plasmodium with a dissecting microscope to observe cytoplasmic streaming. If you are careful, you will see that the direction of streaming is constantly changing. Does the streaming occur in both directions for the same length of time? What do you think is the purpose of these shifts in the protoplasm?

QUESTIONS FOR FURTHER THOUGHT

1. What kinds of adaptations allow the monerans to survive unfavorable environmental conditions? Do protists have similar adaptations?

2. What is the difference between a facultative anaerobe and an obligate anaerobe?

3. How does anaerobic photosynthesis differ from aerobic photosynthesis?

4. How does the requirement for locomotion influence the general physical shape of an animal? Compare the basic design of fast-moving and nonmotile animals. Does *Stentor* change shape when it stops swimming?

5. How are diatoms important in our lives?

TOPIC **24** THE PLANT KINGDOM

Biological Science, 4th ed.
Chap. 40.
Elements of Biological Science, 3rd ed.
Chap. 31.

The plant kingdom may be separated into two basic groups: the Thallophyta and the Embryophyta. The thallophytes include the green, brown, and red algae, while the terrestrial plants are included in the embryophytes. These two groups may be distinguished by the amount of tissue differentiation, their reproductive structures, and whether or not the development of the embryo takes place within the female reproductive organs.

In this unit we will study representatives of most of the divisions of the plant kingdom. We will also examine some adaptations in those nonvascular plants confined to an aquatic or a very moist habitat, as well as adaptations in vascular plants that allow successful life on land.

DIVISION CHLOROPHYTA (GREEN ALGAE)

The green algae are of special interest because they exhibit such a wide variety of forms and because comparisons of their biochemistry with that of the vascular plants strongly implies that these algae were ancestral to terrestrial plants.

Chlamydomonas One of the simplest green algae—probably much like that from which more complex multicellular forms arose—is the unicellular *Chlamydomonas* (Figure 24-1). Its internal structure is usually visible only under very high magnification (1,000× and up), but swimming individuals can be seen clearly under low and high powers. Make a wet mount for observation of their locomotion.

Life cycle   *Chlamydomonas* is haploid. It reproduces asexually by the mitotic production of **zoospores** and sexually by the mitotic production of **gametes**. Vegetative cells, zoospores, and gametes are practically indistinguishable; note that both types of reproductive cells are produced by mitosis.

Gametes fuse to form a **zygote,** the only diploid stage in the life cycle. The zygote immediately undergoes meiosis. The cells that result are freed and eventually develop into new vegetative plants. Figure 24-2 shows a schematic diagram of the life cycle, for comparison with life cycles of other plants.

Sexual reproduction: isogamy   Sexual reproduction probably originated through the union of haploid cells of the same size and structure, a condition called **isogamy.**

Isogamy can be demonstrated in the laboratory by using specially prepared cultures of opposite mating strains of *Chlamydomonas.* Without con-

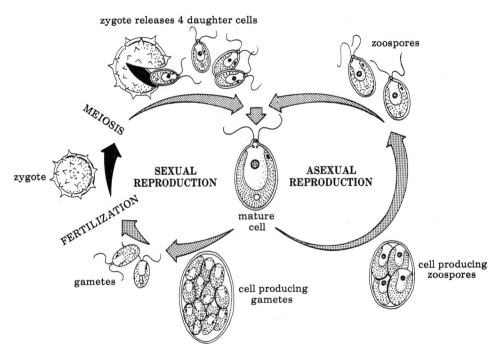

Fig. 24-1. Life history of *Chlamydomonas*. This diagram shows all stages of both the sexual and the asexual cycle. Note the large chloroplast in the central vegetative cell.

taminating either dropper with the opposite culture, obtain single drops of each culture (marked + and −), and allow them to mingle on a slide. Add a cover slip, and examine the slide under low and high powers.

Many of the gametes clump together. Shortly thereafter, they leave the clumps in pairs, fused together at the flagellar ends. Close observation will reveal numerous stages in fusion after the pairing off.

Your instructor may provide a film of sexual reproduction in *Chlamydomonas*.

Heterogamy and the Division of Labor   The green algae we shall examine in this section display two important trends. One is a tendency toward

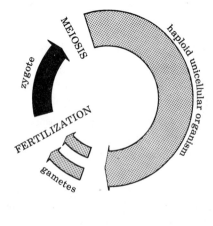

haploid stages
diploid stages

Fig. 24-2. Schematic diagram of *Chlamydomonas* life cycle for comparison with that of other organisms. Note that the zygote is the only diploid stage. This type of life cycle was probably characteristic of the first sexually reproducing unicellular organisms, and it may be the type from which all other types arose.

heterogamy, sexual reproduction that involves interaction between clearly different gametes. The difference can be of size or motility. The most extreme form of heterogamy—in which the female gametes are large, nonmotile eggs and the male gametes are small, motile sperms—is called **oogamy**.

The other tendency involves the organization of cells into colonies or multicellular plant bodies, with associations among the cells and consequent division of labor; cells become specialized for particular tasks.

The volvocine series

There are several green algal genera which, taken together, form a series of colonies of increasing size, internal differentiation, and tendency toward heterogamy. Individual cells of these colonies are much like *Chlamydomonas.* We shall examine *Pandorina* and *Volvox,* two members of what is called the volvocine series (because of its terminal member, *Volvox*). Your instructor may supply you with other members of the series.

1. Make a wet mount or examine a prepared slide of *Pandorina.* Colonies vary in size from 4 to 32 cells; in the larger colonies there should be a clear difference in cell size at the "anterior" and "posterior" ends. If you have living colonies, watch the movement. Does one end move consistently forward? Which end has the larger cells?

Asexual reproduction in *Pandorina* is much like that in *Chlamydomonas;* sexual reproduction, however, is heterogamous: the gametes are of different size, though both are flagellated.

2. Make a wet mount of *Volvox,* or examine a prepared slide of sexual-reproductive stages. Do both, if possible. *Volvox* colonies (or should they be called multicellular organisms?) vary in size from several hundred to as many as 40,000 cells arranged in the wall of a hollow sphere and interconnected by cytoplasmic strands.

Asexual reproduction occurs through the formation of **daughter colonies,** which develop mitotically from single vegetative cells and are eventually freed into the interior of the parent sphere. If you have living colonies, look for some with daughter colonies in the interior of the parent sphere. What do the parental cells look like? When the parent colony dies and breaks apart, the daughter colonies are freed. You may be able to find parent colonies in various stages of degeneration, with daughter colonies inside. What do you suppose initiates the breakdown of the parent colony?

Sexual reproduction is oogamous; eggs and sperms are formed (Figure 24-3). *Volvox* shows a marked division of labor: only a few of the cells are active in gamete production, while the rest are permanently vegetative.

Multicellular algae:
Ulothrix

Make a wet mount or examine a prepared slide of *Ulothrix.* Though relatively simple in appearance, this organism is multicellular and not colonial; adjacent cells share common end walls, and there is some cellular differentiation. Locate the nucleus and single large chloroplast in vegetative cells; on a whole filament, find the **holdfast** cell, which attaches the organism to some substrate.

Any cell of the filament may reproduce asexually by becoming a **sporangium** (a spore-producing structure) and producing zoospores, each of which has four flagella. The zoospores may swim about for some time, then settle, attach, and grow into a new filamentous **thallus** (plant body). You have seen that zoospores can be produced by *Chlamydomonas* and by *Ulo-*

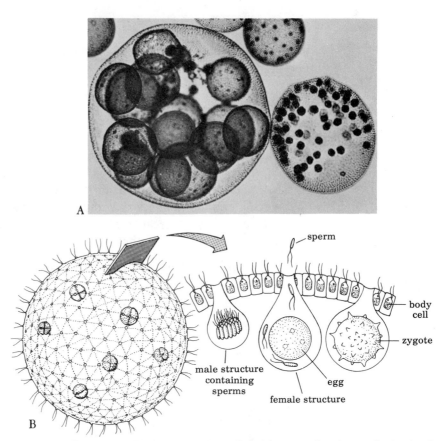

Fig. 24-3. *Volvox.* A: Living vegetative cells with many daughter colonies in the interior of the parent colony; B: section through the surface of a colony, showing male and female reproductive structures. Sperms released by the male structures enter the female structure and fertilize the egg. After a period of inactivity, the zygote divides meiotically, and the haploid cells thus formed then divide mitotically, producing a new daughter colony, which is eventually released. *(A: Courtesy Carolina Biological Supply Company; B: Modified from H. J. Fuller and O. Tippo, College Botany, Holt, 1949.)*

thrix; zoospores are a common method of asexual reproduction among the algae. What value would a *motile* asexual phase have for an attached algal species like *Ulothrix?*

Sexual reproduction is isogamous. The life cycle is shown in Figure 24-4; it is typical of the green algae. As you can see, it differs from that of *Chlamydomonas* only in having a multicellular haploid phase. The life cycle is schematically diagrammed in Figure 24-5.

Spirogyra   *Spirogyra,* an aberrant but fascinating form often studied by beginning students, is a multicellular green alga with a filamentous thallus. Examine a wet mount or a prepared slide of this alga; in vegetative cells, locate the **spiral chloroplast** (studded with numerous **pyrenoids,** starch-production sites) and the large **nucleus,** which is situated in the center of the cell. Cytoplasm runs in strands from around the nucleus to the walls of the cell.

Spirogyra exhibits a peculiar form of heterogamy in which the cells of two filaments of opposite mating strains **conjugate,** or become joined, by

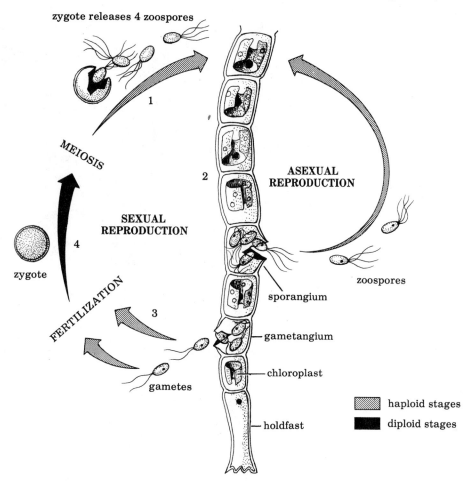

zygote releases 4 zoospores

MEIOSIS

1

2

ASEXUAL
REPRODUCTION

SEXUAL
REPRODUCTION

4

zygote

FERTILIZATION

3

zoospores

sporangium

gametangium

chloroplast

gametes

holdfast

haploid stages

diploid stages

Fig. 24-4. Life cycle of *Ulothrix*. The haploid plant may reproduce either asexually or sexually, though a single filament would never reproduce both ways simultaneously, as shown here. Asexual reproduction is more common; certain cells of the filament develop into sporangia (spore-producing structures) and produce zoospores, which settle down and develop into new filaments. Under certain environmental conditions, the filaments may cease reproducing asexually and begin reproducing sexually; a cell becomes specialized as a gametangium (gamete-producing structure) and produces isogametes. Two such gametes may fuse in fertilization, producing a zygote, which divides meiotically and releases zoospores. What difference would there be between asexually and sexually produced zoospores?

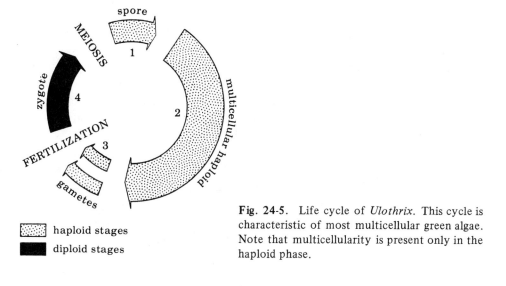

spore

MEIOSIS

1

zygote

4

2

multicellular haploid

FERTILIZATION

3

gametes

haploid stages

diploid stages

Fig. 24-5. Life cycle of *Ulothrix*. This cycle is characteristic of most multicellular green algae. Note that multicellularity is present only in the haploid phase.

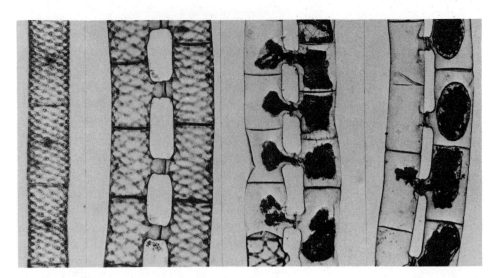

Fig. 24-6. *Photomicrograph of vegetative and conjugating cells of* Spirogyra. *(Courtesy Carolina Biological Supply Company.)*

outgrowths of adjacent cells (Figure 24-6). The entire cell body of each partner cell is then transformed into an amoeboid gamete. Although the gametes look alike, only one of them moves through the tube formed at conjugation. After the migration, the nuclei fuse, a zygote is formed, and, later, meiosis takes place. Three of the four nuclei produced by meiosis degenerate, and the fourth forms a new vegetative filament. How would this degeneration of nuclei be adaptive to the plant?

If your slide shows conjugation, locate the stages in formation of the conjugation tube, formation of gametes, migration of gametes, and formation of zygote. Label the photomicrograph; sketch, adjacent to it, intermediate stages you can locate on your slide.

Ulva You have seen two routes taken by green algae in the development of multicellularity: formation of motile spheres in the volvocine series and formation of filaments in *Spirogyra* and *Ulothrix*. A third route results in the formation of nonmotile associations of cells with platelike thalluses. Since these algae usually live and are attached in areas of relatively rough water, there is often differentiation of blade, stalk, and holdfast.

Examine live *Ulva* (sea lettuce) or *Enteromorpha,* a close relative of *Ulva.* Obtain a small piece of the blade, make a wet mount, and observe the arrangement of cells.

The *Ulva* life cycle (Figure 24-7) differs from that of the other genera considered thus far in that it has a multicellular diploid phase. While the other sexually reproducing organisms you have seen exhibit alternation between haploid and diploid phases, only one of the phases ever becomes multicellular. In many of the more advanced algae, however, both phases are multicellular. The multicellular *haploid* phase is called a **gametophyte** because it produces gametes; the multicellular *diploid* phase is called a **sporophyte** because it produces spores. Such a life cycle is said to exhibit **alternation of generations**. (*Note:* The terms "gametophyte" and "sporophyte," though widely applied, are properly used only when there is alternation of *multicellular* generations.)

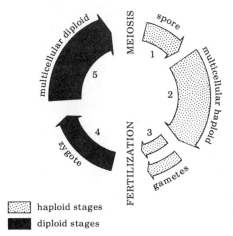

haploid stages

diploid stages

Fig. 24-7. Life cycle of *Ulva*. The gametophyte (multicellular haploid stage) and sporophyte (multicellular diploid stage) are equally prominent.

DIVISION PHAEOPHYTA (BROWN ALGAE)

Laminaria   Examine a demonstration specimen of *Laminaria* or one of the other kelps; the thallus is a sporophyte. The gametophyte is much smaller; sexual reproduction is oogamous. In the sporophyte, notice the rootlike holdfast, stemlike **stipe**, and expanded leaflike blade. The cellular construction of these parts is much more complex than in an alga like *Ulva;* the stipe of some kelps has an outer surface tissue (epidermis), a middle tissue (cortex) containing many plastids, and a central core tissue. Some kelps have a meristematic layer similar to the cambium of the higher vascular plants, and, in a few species, there is a phloemlike conductive tissue in the central core. In short, these brown algae are complex plants that have convergently evolved many striking similarities to the vascular plants.

Fucus   *Fucus,* a common brown alga in northern coastal waters, is unusual among plants in that it has a life cycle that omits the multicellular haploid phase; gametes are, therefore, formed by meiosis directly from the diploid phase. In this respect, *Fucus* and its immediate relatives have a life cycle closely comparable to that of animals.

1. Examine a live *Fucus,* if possible. Notice the holdfast, if it is still present, and its firm attachment to rock. The thallus bears **bladders** (what is their function?) and, at the tips of fertile plants, **receptacles** (Figure 24-8).

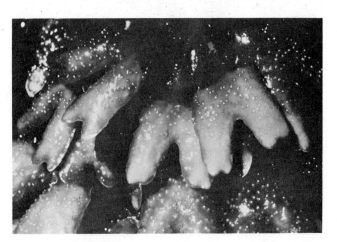

Fig. 24-8. Thallus of *Fucus,* bearing receptacles. *(Courtesy Carolina Biological Supply Company.)*

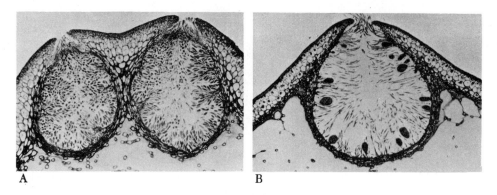

Fig. 24-9. *Fucus* conceptacles. A: Male; B: female. *(Courtesy Carolina Biological Supply Company.)*

Examine the receptacles under a dissecting microscope, and note the numerous tiny openings. Each opening leads into a cavity, called a **conceptacle**, where the sperms or eggs are located.

2. Make hand sections of male and female receptacles, and locate thin sections of conceptacles. (Your instructor will distinguish the male and the female plants for you if the species you are using is dioecious; if it is monoecious, section receptacles until you find conceptacles comparable to Figure 24-9.) Notice that most of the cells in the conceptacle are sterile, colorless filaments. Terminal cells on some of the filaments become fertile: in male conceptacles they are called **antheridia**, and in females, **oogonia**. Many sperms develop from each antheridium and several eggs from each oogonium. Gametes are extruded from the receptacles of fertile plants, and fertilization occurs externally. The zygotes settle, attach, and develop into new diploid plants. The life cycle is diagrammed schematically in Figure 24-10.

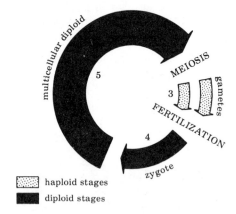

haploid stages
diploid stages

Fig. 24-10. Schematic life cycle of *Fucus*.

3. If the material available to you is not fertile, examine prepared slides of sections through male and female conceptacles. The instructor will prepare a demonstration of fertilization in *Fucus* if he or she is able to obtain live fertile material.

4. If you have time, make hand sections across various parts of the blade to study its internal structure. The plant body is composed of masses of branching filaments, most of which are colorless. Only the outermost filaments are photosynthetic.

DIVISION BRYOPHYTA (MOSSES AND LIVERWORTS)

It is little wonder that so many organisms live in the water for most or all of their lives, for water is more secure and stable than any other habitat. The terrestrial environment, in contrast, is harsh and insecure, constantly changing, and, therefore, hazardous and hostile to most living organisms. It is not surprising that its successful inhabitants have diverse and often complex adaptations that enable them to cope with the trials of terrestrial existence. Many of the attributes of terrestrial organisms cannot be understood properly without some knowledge of the problems and challenges they must face and meet in order to survive.

The difficulties facing a terrestrial organism are numerous. It must be able to obtain an adequate supply of water, both when water is abundant in the environment and when it is not. It must have a means of physical support, unnecessary in buoyant water; and, if it is an animal, it must be able to move about effectively. It must reproduce in an environment that seldom provides sufficient water for flagellated sperms to swim in and offers little security for the developing zygote. Finally, a terrestrial organism must be able to withstand many changes in temperature, humidity, wind, and light, and must be able to adapt to the other environmental fluctuations to which it is frequently subject.

All the modern multicellular photosynthetic plants that spend any part of their life cycle on land—including plants as diverse as mosses and trees—enclose their zygotes and early embryos in a protective jacket of cells and are, therefore, grouped together in the category Embryophyta. The Thallophyta, in contrast, are characterized by exposed embryos or by embryonic development entirely apart from the parent plants.

We have little information about the ancestors of terrestrial plants. Largely on the basis of biochemical similarities, the algal division Chlorophyta is supposed to have been ancestral to land plants, but we have only scant fossil evidence to suggest the manner in which a transition might have taken place. The least complex living Embryophyta, and the only nonvascular members of that group, are the mosses, hornworts, and liverworts, of the division Bryophyta.

Vegetative structure of moss

Examine an individual moss plant under the dissecting microscope. The green, "leafy" plant, familiar among the flora of water's edge, moist rock, and woodland, is the multicellular haploid phase—the gametophyte. Note the stemlike axis of the plant and its leaflike appendages. The aerial parts of the plant are cuticle-covered, an important adaptation to terrestrial life. Remove a "leaf," and examine it in a wet mount. Are there guard cells and stomata?

The gametophyte is anchored in the soil and obtains water by means of **rhizoids**, outgrowths at the base of the axis. Moss gametophytes lack vascular tissues, which sharply limits their size and the sort of habitat they can occupy.

Sexual reproduction in moss

1. The distribution of mosses is limited to moist places because mosses require water for sexual reproduction. Male organs (**antheridia**) and female organs (**archegonia**) occur, in season, at the tips of the gametophyte plants. Examine slides of longitudinal sections through these regions, and locate the

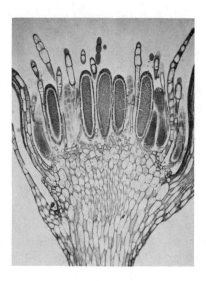

Fig. 24-11. Longitudinal section through the antheridium of *Mnium* moss. *(Courtesy Ward's Natural Science Establishment, Inc., Rochester, New York.)*

sexual organs. Antheridia are club-shaped and are filled with developing sperms; archegonia are vase-shaped, and in perfect longitudinal sections you will be able to find an egg at the base of the archegonium. Many sterile filaments occur among the sexual organs in both sections. Label Figure 24-11.

Flagellated sperms, released from the antheridia, swim through the dew that frequently covers the gametophytes, or splash in drops of water from the male to the female plant. Once on the female plants, sperms are attracted to the eggs by chemical exudates from the archegonia. Fertilization occurs in the base of the archegonium. The zygote, deriving nutrient from the female plant, gives rise mitotically to a capsule-tipped stalk. Meiosis occurs in the capsule, forming thousands of spores. Stalk and capsule together constitute the diploid sporophyte.

2. Examine the sporophyte of one of the specimens. If the capsule has not already released its spores, carefully remove the top of the capsule, and examine the interior with a dissecting microscope. Just below its apex, the capsule bears a ring of tiny teeth that, with periodic changes in humidity, bend inward and outward, dispersing the spores.

How would a high-growing sporophyte benefit the moss plant?

3. Spores, when they land in the proper place, germinate into filamentous young gametophytes, structurally reminiscent of the filamentous green algae. Branches of the filaments growing down into the soil become the rhizoids, and branches growing upward develop into the familiar "leafy" axes. Under the dissecting microscope, examine an agar plate upon which moss spores have been germinated. Find and sketch as many developmental stages as you can. A mature moss gametophyte (with attached sporophyte) and a schematized bryophyte life cycle are shown in Figure 24-12.

Vegetative Structure of a Liverwort Liverworts occur in habitats much like those of moss except that the former require more water. Examine vegetative plants of *Marchantia,* using a dissecting microscope and making hand sections. Examine prepared slides, if available. Is the whole

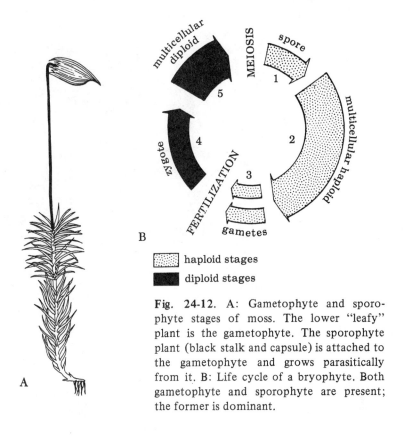

Fig. 24-12. A: Gametophyte and sporophyte stages of moss. The lower "leafy" plant is the gametophyte. The sporophyte plant (black stalk and capsule) is attached to the gametophyte and grows parasitically from it. B: Life cycle of a bryophyte. Both gametophyte and sporophyte are present; the former is dominant.

Marchantia thallus photosynthetic? What function do pores in the top of the thallus serve? Compare the adaptations to terrestrial life of *Marchantia* with those of the moss.

Sexual Reproduction in Liverworts   The sexual cycle in liverworts is almost exactly like that of moss except that antheridia and archegonia are borne on special aerial portions of the gametophyte plants. The sexual organs themselves are much like those of moss.

Examine the antheridium-bearing part of a gametophyte, comparing it to Figure 24-13. Examine the archegonium-bearing part of a gametophyte,

Fig. 24-13. Thallus of *Marchantia*, with antheridium-bearing stalks. *(Courtesy Carolina Biological Supply Company.)*

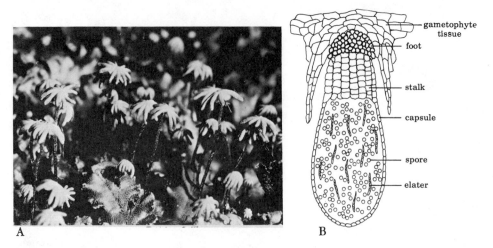

Fig. 24-14. *Marchantia.* A: Thallus with archegonium-bearing stalks. One of the mature stalks, at the arrow, is bearing young sporophytes on its undersurface. B: *Marchantia* sporophyte with foot embedded in gametophyte tissue, short stalk, and capsule. The capsule contains spores and elaters, the latter being elongate structures that aid in dispersal of the spores. *(A: Courtesy Carolina Biological Supply Company.)*

comparing it to Figure 24-14. As in moss, the zygote formed at fertilization (where?) develops into the sporophyte; the liverwort sporophyte hangs downward from the archegonium-bearing part of the gametophyte. How are the spores dispersed?

If available, be sure to see a film of sperm transfer and fertilization in *Marchantia.*

FERNS (SUBDIVISION PTEROPSIDA)

Like most vascular plants, ferns exhibit an alternation of multicellular generations in which the diploid multicellular sporophyte is the dominant phase.

Structure of the
fern sporophyte

1. Examine a mature fern sporophyte, the familiar leafy woodland plant. Locate the **rhizome** (an underground stem) and the **fronds,** each of which is a complexly subdivided leaf. Are the fronds cuticle-covered? Do they have stomata and guard cells? If sufficient material is available, make hand sections of frond axis and of rhizome, and stain the sections with phloroglucinol-HCl. (If these materials are unavailable, examine prepared slides.) Examine the slides microscopically. Phloroglucinol-HCl stains lignin a bright red. (Lignin is a cellulose derivative that occurs in the walls of xylem vessels, sclerenchyma, and other strengthening cells.) Notice the difference between the arrangement of vascular tissues in ferns and in flowering plants, examined in Topic 7. Is secondary growth possible in a fern? How does the sporophyte survive the changing seasons?

Using your general knowledge of stem structure and your observations of rhizome sections, label Figure 24-15.

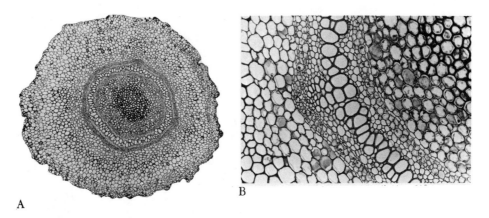

A   B

Fig. 24-15. Cross sections through fern rhizome. A: Low power, showing whole rhizome; B: high power, showing a portion of the vascular region. *(Courtesy George H. Conant, Triarch Incorporated, Ripon, Wisconsin.)*

2. On the underside of some or all of the fronds, notice brownish **sori** (singular, "sorus"); each one is a cluster of many **sporangia**. The structure of the sorus varies from species to species, but it often includes an umbrella or shelflike structure, the **indusium,** that protects the sporangia until they are mature. In Figure 24-16, locate and label sporangia and indusium. Remove a small sorus-bearing region from a frond, examine it with a dissecting microscope, and then remove some of the sporangia for closer microscopic observation.

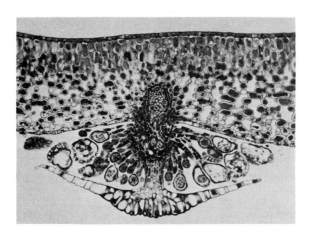

Fig. 24-16. Fern leaf and sorus, median longitudinal section. *(Courtesy Ward's Natural Science Establishment, Inc., Rochester, New York.)*

Each sporangium is crested by a special ring of cells (the **annulus**) with unevenly thickened walls. When a mature sporangium dries, the cells of the annulus lose water and their walls shrink; this results in the bending of the annulus as shown in Figure 24-17. Eventually the tension of the small amount of liquid water in the annulus cells becomes greater than the cohesiveness of water. The water turns instantly from liquid to vapor, restoring the annulus cells to their former size and shape and causing a violent snap of the annulus back to its former position. This movement is generally effective in dispersing the spores a great distance. If you have mature but unopened sporangia with spores clearly visible inside, you may be able to see some of

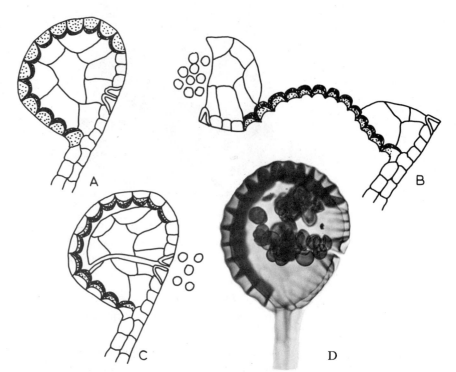

Fig. 24-17. Ejection of spores by a fern sporangium. A: An intact sporangium; B: top of a sporangium bent back as a result of loss of water from cells of the annulus; C: top of the sporangium snaps shut, releasing spores; D: photograph of a sporangium, showing the annulus. *(A–C: Modified from F. C. Steward,* About Plants: Topics in Plant Biology, *Addison-Wesley, 1966. Used by permission. D: Courtesy Carolina Biological Supply Company.)*

the dramatic annular movements by allowing the sporangia to dry under the heat of an incandescent lamp.

3. Spore-bearing fronds are called **sporophylls** (*sporo-,* "spore"; *-phyll,* "leaf"). In many ferns, all the fronds become sporophylls; in others, only a few fronds become sporophylls. In some species the fronds that become sporophylls are so highly modified at maturity that they do not even resemble vegetative fronds at all. Into which of these categories does your specimen fit? Examine other types on demonstration. We shall see that in angiosperms and conifers, the most fully successful terrestrial plants, spore-bearing is always restricted to groups of a few highly modified leaves—flowers or cones—that bear little resemblance to foliage leaves.

The fern gametophyte

1. Fern spores are produced by meiosis in the sporangium and are dispersed by the wind. Spores that land in favorably moist environments germinate and undergo mitosis to form a heart-shaped multicellular gametophyte called a **prothallus**.

With a dissecting microscope, examine specimens of immature and sexually mature fern prothalli. Note their size and general shape; in the older, sexually mature specimens, notice prominent **rhizoids**. If enough living material is available, make wet mounts of the mature prothalli, and

archegonium

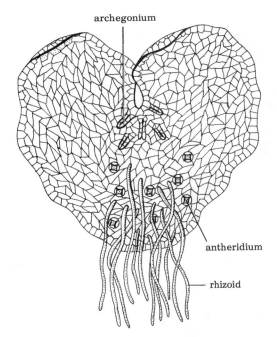

antheridium

rhizoid

Fig. 24-18. Fern gametophyte. This is a highly magnified view of the underside of the prothallus. *(Modified from H. J. Fuller and O. Tippo,* College Botany, *Holt, 1949.)*

locate the **antheridia** and **archegonia** (much like the sexual organs of moss and liverwort), borne on the underside of the prothalli (Figure 24-18).

2. Ferns are **homosporous**—i.e., they produce only one morphological kind of spore. Though the spores may be genetically variable (why?), each one gives rise to a haploid gametophyte that bears both antheridia and archegonia, and the gametes these organs produce are genotypically identical, since they are produced by mitosis. It would be advantageous to the plants to discourage self-fertilization and encourage union between sperms from one gametophyte and eggs from another. By what mechanisms could the plant accomplish this? Considering the fact that self-fertilization defeats part of the selective value of sexual reproduction, can you suggest any advantages to homospory and the consequent production of both sexual organs on the same gametophyte?

In our examination of the reproduction of the higher vascular plants, we shall see that any advantages of homospory have been sharply offset by the much greater advantages of **heterospory**—production of different "male" and "female" spores, and, hence, different gametophytes.

3. The fern zygote (where is it formed?) develops into a sporophyte that quickly becomes independent of the gametophyte; the prothallus soon disintegrates. Examine, on demonstration, an old prothallus with a young sporophyte attached.

From one point of view, ferns are no better adapted to terrestrial life than are the bryophytes. Even though the vascularized sporophyte can attain a larger stature, the fragile gametophyte can survive only in moist places. Remember, though, that sexual organs are formed on the underside of the prothallus, and that the gametophyte, though fragile, is also small and can grow close to the ground where it is not as susceptible to drying. Do you think the size and structure of the fern gametophyte would encourage the dispersal of ferns to and their survival in habitats not open to mosses and

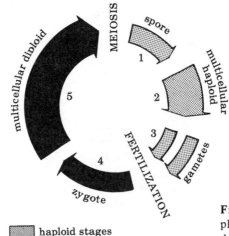

haploid stages
diploid stages

Fig. 24-19. Life cycle of ferns. Both gameto-phyte and sporophyte are present; the latter is dominant. Compare this life cycle with that of a bryophyte (Figure 24-12B).

liverworts? Would a large mosslike gametophyte have the same advantages as a fern gametophyte?

The Fern Life Cycle The fern life cycle is diagrammed schematically in Figure 24-19, for comparison with other plant life cycles. We have noticed in ferns several trends that have been sharply accentuated in higher vascular plants:

1. The diploid sporophyte is the dominant phase in the life cycle, and the haploid gametophyte is much reduced in size. Though still multicellular in all terrestrial plants, the gametophyte in flowering and cone-bearing plants is an inconspicuous phase in the life cycle; it is not free-living and photo-synthetic, as it is in ferns and mosses and liverworts.

2. Certain leaves of the sporophyte may become specialized for spore-bearing, while other leaves remain vegetative. This trend begins in the ferns and is carried to extremes in the flowering and cone-bearing plants, where a few groups of leaves are highly modified to bear spores. While ferns are homosporous, heterospory is the rule in the more fully terrestrial seed plants.

SEED PLANTS (SUBDIVISION SPERMOPSIDA)

In Topics 4 and 7, we considered the structural basis of water and mineral procurement, gas exchange, and transport in seed plants. In doing so, we reviewed many of the adaptations that have enabled these plants to attain their present dominant and successful position among land flora. We saw, for example, that a complex vascular system carries materials throughout the plant; water and minerals are absorbed by the roots and carried upward to the stem and leaves in the xylem, while the products of photosynthesis are carried to other parts of the plant by the phloem. We saw that guard cells, interrupting the cuticle-covered leaf epidermis, serve to control the diffusional loss of water from stomata, and that natural selection has turned this loss (transpiration) to advantage in moving materials in the xylem. We saw that both stem and root have been strengthened by the presence of cells with secondarily thickened walls (such as sclerenchyma) and that some of the secondarily thickened cells (such as xylem vessels and tracheids) serve the double function of support and transport, permitting a greater stature than is possible in nonvascular plants. Finally, we saw that secondary growth, by adding to the girth of stem and root, makes possible a larger plant body and provides the strength that often carries such plants through severe climatic stress.

In this section, we shall study the reproductive adaptations that have contributed to the dispersal of seed plants and to their survival in a wide variety of terrestrial habitats. We shall confine our study to two groups of seed plants: the conifers (class Coniferae) and the flowering plants (class Angiospermae).

Conifers   Hemlock, pine, and spruce are conifers with which you are probably familiar. In each of these, the large green tree is the sporophyte, which is the dominant phase in the life cycle.

To put reproductive adaptations of seed plants in their proper perspective, we shall summarize some of the salient differences between the reproduction of ferns and that of the seed plants.

• Like ferns, the seed-plant sporophyte is the dominant phase in the life cycle; the gametophyte, while present, is reduced.

• In ferns, we noticed a tendency to restrict spore-bearing to special leaves (sporophylls). In seed plants, this tendency is more evident; the spore-bearing parts of the plant are cones (in conifers) or flowers (in flowering plants). These structures represent groups of highly modified sporophylls.

• Unlike ferns, which exhibit homospory, the seed plants produce different "male" and "female" spores, often on separate parts of a plant or on separate plants. The "male" spores are often called **microspores** and the "female" spores **megaspores**.

• While fern spores develop into the haploid multicellular gametophyte —a green, free-living structure—seed-plant gametophytes are not green and never become free-living. Instead, gametophyte development occurs on the parent sporophyte. The male gametophytes (pollen grains) are generally carried to the female gametophyte by wind or by animals. Fertilization and the development of seeds occur on the parent sporophyte plant. The seeds are then released to "fend for themselves."

Examining a conifer cone   1. Examine a pine or other conifer branch with **cones** attached. Locate the compact green female cones and the smaller, lighter male cones. (Large green or brown cones are the older female cones. Ignore them for the moment.) Both male and female cones are composed of **cone scales** (each one a sporophyll) arranged on a short section of stem. Each scale bears one or more sporangia within which haploid spores are produced by meiosis.

In the male cone, each haploid spore develops into a four-celled pollen grain—the male gametophyte, Figure 24-20.

In the female cone, each scale bears two sporangia, each enclosed in a protective integument except for a small opening, the **micropyle** (Figure

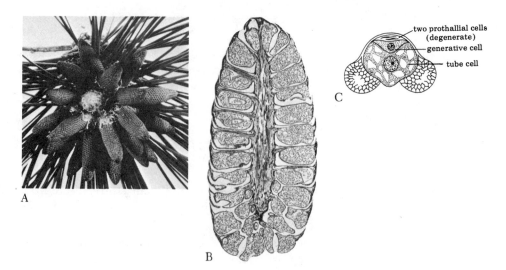

Fig. 24-20. Male pine cones. A: Mature male cones. B: Longitudinal section through a male cone. Each sporophyll (cone scale) bears a large sporangium that becomes a pollen sac. C: Pollen grain (male gametophyte) composed of four cells, two of which are degenerate. *(A, B: Courtesy Carolina Biological Supply Company.)*

24-21). Within the sporangium, meiosis occurs, forming four haploid megaspores. Three of these disintegrate, and the fourth divides mitotically to form a multicellular mass, the female gametophyte. When mature, the female gametophyte produces two to five archegonia at its micropylar end; egg cells develop in each archegonium. Each sporangium, together with the gametophyte inside it and the integument surrounding it, is called an **ovule**.

2. When the male, or pollen, cones ripen, the sporangia open and the pollen is released and carried away by the wind. If available, examine microscopically some pine or spruce pollen. Notice that each grain has winglike processes that probably assist in its dispersal. Each pollen grain is covered with a protective cuticle and is thus protected against water loss.

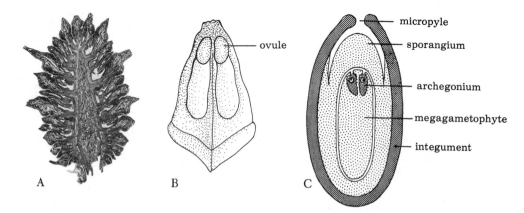

Fig. 24-21. Female pine cones. A: Longitudinal section through a female cone. Ovules can be seen on the surface of the sporophylls, near their base. B: Scale from a female cone. The two ovules lie on the surface of the scale, near its point of attachment to the cone axis. C: Section through an ovule, showing the female gametophyte. *(A: Courtesy Carolina Biological Supply Company; B, C: Modified from H. J. Fuller and O. Tippo, College Botany, Holt, 1949.)*

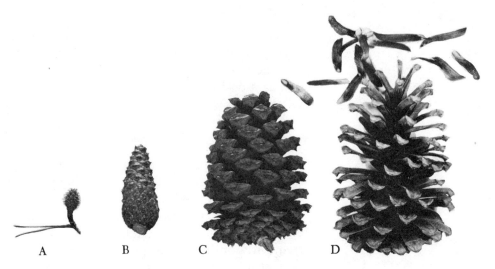

Fig. 24-22. Maturation of female pine cones. A: Female cone in the first year of growth; B: in the second year; C: in the third year; D: female cone with seeds. *(Courtesy Carolina Biological Supply Company.)*

Most of the pollen never reaches the female gametophyte, but some of it shifts down between the female cone scales and comes into contact with a sticky secretion at the micropyle of an ovule. As the secretion dries, the pollen grain is drawn into contact with the sporangium and promptly begins to germinate, developing a tubular outgrowth, the **pollen tube,** which grows through the tissue of the sporangium and penetrates the female gametophyte.

One haploid nucleus, derived from one of the pollen-tube nuclei, fertilizes the egg when the pollen tube reaches the archegonium. The resulting zygote develops into an embryo, which, along with food material derived from the female gametophyte, is enclosed in a protective coating derived from the old integument of the ovule. The entire structure is shed from the surface of the cone scale as a **seed.**

3. Development from pollination to shedding of the seed may take several years in some conifers. During this period, the female cone increases greatly in size. On a tree, you should be able to locate several different stages in female-cone development (see Figure 24-22). In maturity, the female cone dries, the scales separate from one another, and the seeds fall out. This mature cone is the one often used in Christmas decorations.

If sufficient material is available or if you are examining cones on a tree in the field, remove cones in several stages of development, and carefully dissect the scales. In the youngest cones, you will find the ovules in pairs at the base of each scale; in developing cones, you may find ripening seeds; and in mature cones, you will probably find a substantial number of scales from which seeds have already fallen. Under the dissecting microscope, dissect a seed. Locate the embryo and stored food material.

Flowering Plants While flowers and the plants that produce them are familiar to us all, it is not readily apparent that the flower, like the cone, is actually a group of spore-bearing leaves (sporophylls) so highly modified for the production and protection of male and female gametophytes that they no longer bear any resemblance to foliage leaves. Flowers thus carry to extremes a tendency to restrict spore-bearing to specialized leaves. We saw a few ferns that displayed this tendency, and we encountered it again, in much more pronounced form, in conifers. The manner in which flowering plants bear their sporangia is fundamental to their present highly successful and dominant position on land.

While flowering plants have a life cycle much like that of conifers, let us review some differences between these groups before examining in detail the reproduction of flowering plants.

● While conifer ovules are borne on the surfaces of highly modified leaves (cone scales), ovules of flowering plants are borne within a group of even more highly modified leaves in a section of the flower called the ovary.

● While conifer pollen reaches conifer ovules principally on wind currents, the pollen of flowering plants is often carried by some animal agent (frequently an insect). Both the animal vectors and the flowers have developed striking adaptations that reflect their intimate relationships. Many flowering plants could not reproduce sexually without their animal pollinizers.

● Finally, the ovary or ovaries of the flower develop into a fruit or fruits, protecting the seeds and often facilitating the species' dispersal.

Examining a typical flower

1. Examine a simple typical flower (e.g., gladiolus), and locate the four sets of modified leaves (Figure 24-23):

    a. **Sepals,** which enclose and protect all the other floral parts during the bud stage. They are usually small, green, and leaflike, but in some species they are large and brightly colored. All the sepals together form the **calyx.**

    b. **Petals,** internal to the sepals and often showy in animal-pollinated flowers. Together, the petals form the **corolla.** In some flowers, petals are reduced or absent altogether. How are these flowers pollinated?

    c. The **stamens,** the male sporophylls, located just inside the circle of the corolla. Each stamen consists of a **filament** and a terminal ovoid **anther,** in which pollen is formed.

    d. The **pistil,** representing several fused female sporophylls. Each pistil consists of an **ovary** at its base, a slender **style** (sometimes

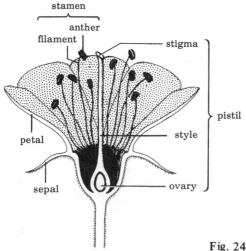

Fig. 24-23. The parts of a flower.

more than one) that rises from the ovary, and an enlarged apex called the **stigma**. That the pistil represents several fused sporophylls in many species is usually evident from the multiple structure of the stigma and from the microscopic structure of the ovary, which is similarly divided. Make cross sections of the ovary of your flower. Notice its subdivided interior, and locate ovules (sporangia) attached to the wall of the ovary by short stalks.

Make similar sections of an anther, and examine microscopically the anther section and some of the pollen.

Flowers having all four sets of modified leaves are called *complete flowers;* flowers lacking one or more of them are called *incomplete flowers.* Incomplete flowers lacking one of the sets of reproductive organs are called *imperfect flowers,* while flowers with both sets of reproductive organs are called *perfect flowers.* Examine flowers on demonstration. Into which of the above categories does each fit? Identify as many of the sets of modified leaves as are present on each, and determine, if possible, what kind of agent and (if an animal) how large an agent effects pollination.

2. Angiosperm gametophytes are formed in the ovules and anthers in much the same way as in conifers. The female gametophyte, commonly a seven-celled, eight-nucleate structure, is formed from one of four haploid megaspores that result from meiosis in the ovule; the gametophyte is called an **embryo sac**. The male gametophytes, or pollen grains, are produced by meiosis much as in conifers.

Pollen deposited on the stigma develops a tube that grows down into the style, contacts the embryo sac (Figure 24-24), and releases two nuclei. One of these nuclei fuses with an egg cell (or nucleus), and the resulting zygote develops into the embryo. The other pollen nucleus fuses with two or

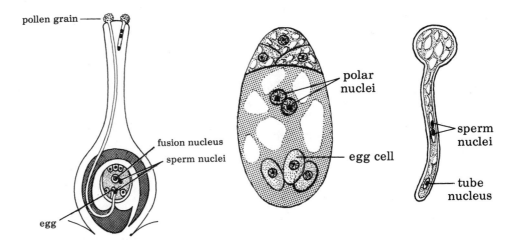

Fig. 24-24. Fertilization in an angiosperm. Pollen grains land on the stigma and develop pollen tubes that grow downward through the style. One of the pollen tubes shown here has reached the ovule in the ovary and has discharged sperm nuclei into it. One sperm nucleus will fertilize the egg cell, and the other will unite with the diploid fusion nucleus (derived from the two polar nuclei) to form a triploid nucleus, which will give rise to endosperm. Details of the female gametophyte (embryo sac) and of the pollen grain are shown in the two drawings on the right. Instead of being exposed, as it is in conifers, the angiosperm female gametophyte is completely enclosed within the ovule inside the ovary of the flower.

more other nuclei in the embryo sac and, by mitotic divisions, gives rise to the endosperm, a tissue that nourishes the developing embryo when the seed breaks dormancy and the young plant resumes growth.

3. The walls of the ovaries of flowering plants develop into the **fruit,** which encloses the **seeds.** The many types of fruit reflect the many different kinds of ovary and a wide variety of adaptations for dispersal. *Simple fruits* are derived from a single ovary; *aggregate fruits* are derived from many separate ovaries of a single flower which later fused to form a single fruit. Examine the seeds and fruits on demonstration. Discuss with your classmates the mechanisms by which these seeds and fruits are dispersed, and defend your suggestions.

Open a simple fruit, such as an apple, by cutting through the center, separating the top half from the bottom half. After examining this cross section, decide how many sporophylls fused to form the ovary.

QUESTIONS FOR FURTHER THOUGHT

1. What advantages are there to sexual reproduction by isogamy? By oogamy?

2. Define alternation of generations, and explain its significance. Which of the following organisms exhibits alternation of generations: *Fucus, Ulva, Spirogyra, Chlamydomonas, Laminaria,* moss, *Homo sapiens*?

3. What evidence suggests that blue-green algae are plants? What evidence suggests that they are not plants?

4. Algae occur in many different colors, some quite unlike the usual green plants. What is the adaptive significance of difference in coloration, particularly in the case of algae that live attached to rocks relatively close to shore?

5. The majority of shore-dwelling algae have holdfasts, differentiations of the plant base that cement the organism to rock or some other substrate. What benefit might the plants gain from such structures? What special adaptations might you expect to find in algae that grew only in mid-ocean?

6. Do holdfasts, or any other structures in algae, function in transport within the plant? Are they involved in nutrient procurement? How does an alga solve the problems of nutrient procurement that are met by the roots of a terrestrial plant?

7. What unique features of bryophytes enable these plants to survive on land? What features make them poorly adapted to land?

8. Describe the association between the moss gametophyte and the moss sporophyte.

9. What asexual reproductive features characterize Bryophyta? Of what advantage is it to terrestrial plants to have both asexual and sexual reproduction?

10. Compare the structure of moss with that of a fern.

11. What is heterospory? What is its significance?

12. With respect to size and independence, compare the gametophyte and sporophyte of a moss with those of a conifer.

13. What are the major features that distinguish conifers from flowering plants?

14. What is a sporophyll? In what groups of plants are sporophylls most strikingly modified? What purpose do the modifications serve? Defend your answers.

15. Where on an angiosperm plant is the gametophyte located?

16. Discuss some of the adaptations by means of which cross-pollination is encouraged in flowering plants.

17. Mendel, in doing his experiments with garden peas, found that peas normally are self-fertilized since maturation of the floral organs occurs before the flower has opened. What procedure would you follow in order to bring about cross-fertilization in such a plant?

TOPIC **25** THE FUNGAL KINGDOM

Biological Science, 4th ed.
Chap. 41, pp. 1095–1102.
Elements of Biological Science, 3rd ed.
Chap. 32, pp. 668–674.

The fungal kingdom is composed of unicellular or filamentous heterotrophic organisms that produce spores. Some fungi are saprophytes and feed on dead organic matter, some are parasites and obtain nutrients from living matter, while others exist in a symbiotic relationship. Fungi are mainly terrestrial organisms, but some marine forms do exist.

The thin, multicellular or multinucleate filaments that make up the fungal organism are called **hyphae**. The mass of hyphae is called a **mycelium**. Fungal cells are enclosed by a cell wall composed primarily of **chitin**, the material that also makes up the exoskeleton of arthropods. Cross walls, or **septa**, form in some of the hyphae, but these usually contain numerous perforations, and cytoplasmic streaming occurs freely throughout the filament. The only time that a completely enclosed cell is produced (when septa form) is during the formation of spores and gametes.

The fungal kingdom is separated into four divisions: Oomycota, Zygomycota, Ascomycota, and Basidiomycota. The distinguishing feature of each of these divisions is the method of sexual reproduction.

In this laboratory, we shall examine the structure and method of reproduction of the four groups of fungi.

OOMYCOTA

These fungi are commonly called "water molds." However, not all Oomycota are aquatic. The hyphae are multinucleate and lack septa until reproductive structures form. The exterior cell walls contain polysaccharides that differ from other fungi, and do not contain chitin. The Oomycota differ in many ways from the other phyla. In the water molds, the mycelium remains diploid through most of the life cycle of the molds rather than haploid as in the other fungi. The Oomycota are the only fungi that reproduce asexually by flagellated zoospores. This is also the only group in which sexual reproduction is oogamous—that is, the egg is nonmotile and large, and the male gametes are small and motile.

Saprolegnia

One of the most commonly encountered members of this division is *Saprolegnia,* also called "fish mold." Make a wet mount of a small piece of the mycelium of *Saprolegnia.* Examine the hyphae for reproductive structures (Figure 25-1). Are these structures for asexual or sexual reproduction? If you cannot locate both types of reproductive structures, examine prepared slides. Do the oogonium and the antheridium contain the same number of nuclei?

262

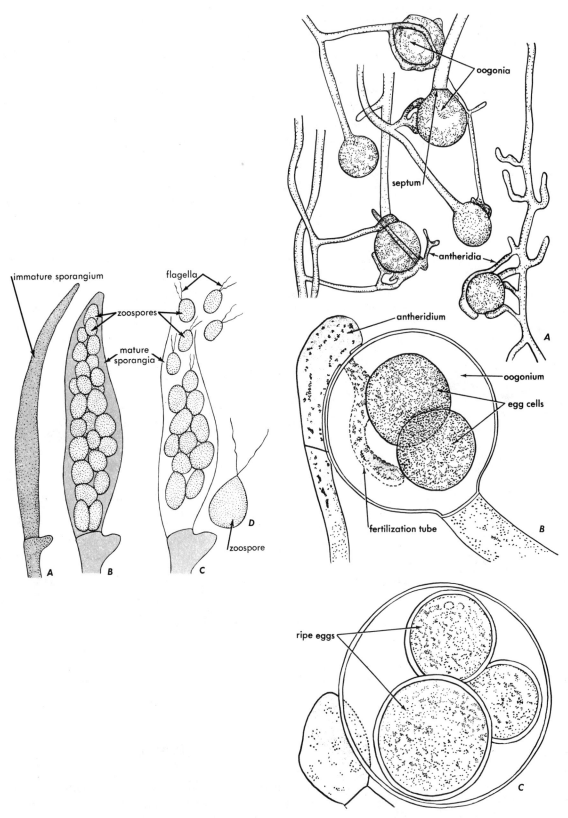

Fig. 25-1. Top: Sporangia of *Saprolegnia.* A: An immature sporangium; B: a mature sporangium; C: discharge of zoospores; D: a zoospore. Right: Gamete-producing structures in *Saprolegnia.* A: Oogonia and antheridia with associated hyphae; B: a more highly magnified view of the antheridium, fertilization tube, and one oogonium bearing two egg cells; C: ripe egg cells. *(Reproduced with permission from Weier, Stocking, and Barbour, Botany, an Introduction to Plant Biology, 6th ed., 1982. Courtesy John Wiley and Sons.)*

ZYGOMYCOTA

The Zygomycota are referred to as the "conjugation fungi." Obtain a culture of the black bread mold *Rhizopus stolonifer.* During sexual reproduction, also referred to as conjugation, hyphae of opposite mating types come in contact and initiate the formation of gametes, which fuse to form a zygote. With a dissecting microscope, locate the three different types of hyphae:

Rhizopus stolonifer

1. Hyphae growing on the surface of the agar.
2. **Rhizoids**, which absorb nutrients and anchor the fungus.
3. Hyphae called **sporangiophores** that grow upright from the surface and bear **sporangia** at their tips. Thousands of asexual spores are produced in the sporangia. Spores are the result of what type of cell division?

Prepare a wet mount of a tiny bit of the mycelium in 0.1% neutral red. Are the cells multinucleate? Are cross walls present in the filaments?

Obtain a petri dish on which plus and minus strains of *Rhizopus* have been inoculated. Note the line of zygotes formed where the hyphae came into contact. Figure 25-2 shows the life cycle of *Rhizopus.* How would cells produced by the zygote differ genetically from those produced in the sporangia?

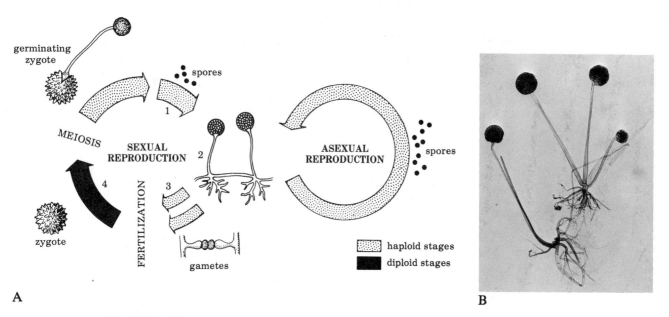

Fig. 25-2. *Rhizopus.* A: Life cycle; B: hyphae with sporangia. *(B: Courtesy Carolina Biological Supply Company.)*

ASCOMYCOTA

The division Ascomycota, also called "sac fungi," includes many diverse forms: unicellular yeasts, morels, truffles, powdery mildews, cottony molds, and complex cup fungi. The distinguishing feature of this phylum is the saclike reproductive structure called an **ascus**, which forms when hyphae of compatible mating types conjugate. In each ascus, eight haploid spores are usually formed (Figure 25-3). The hyphae of the sac fungi possess perforated cross walls.

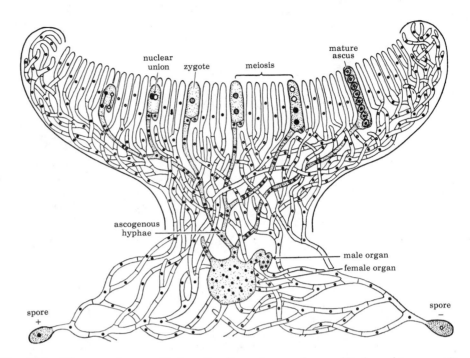

Fig. 25-3. Diagram of a magnified section through a cup fungus. Hyphae of two mycelia, one from a plus (female) spore and the other from a minus (male) spore participate in forming the cup structure and in producing the spores. The plus mycelium bears a female organ, and the minus mycelium bears a male organ. A tube grows from the female organ to the male organ, and then minus nuclei (small circles) acting as male nuclei, move into the female organ and become associated with the female nuclei (black dots). Next, hyphae (gray) grow from the female organ; each cell in these hyphae contains two associated nuclei, one minus and one plus. The terminal cells of these hyphae eventually become elongate, and their nuclei unite, forming a zygote nucleus. The zygote nucleus promptly divides meiotically, and each of the four haploid nuclei thus formed then divides mitotically, producing a total of eight small spore cells, which are still contained within the wall of the old zygote cell, now called an "ascus." When the mature ascus ruptures, the spores are released. *(From L. W. Sharp,* Fundamentals of Cytology, *McGraw-Hill Book Co., 1943. Used by permission.)*

Peziza

Under the dissecting microscope, examine a preserved specimen of *Peziza,* a cup fungus. The cup, called an **ascocarp,** is made up of masses of tightly packed hyphae. Note the inner layer of the specimen. This layer contains the asci, which contain the ascospores. When the ascus is mature, it bursts open and the spores are released and are carried by air, water, or animal.

Obtain a slide of a cross section through the cup. Note the masses of hyphae in the cup. Study the arrangement of the asci and the sterile filaments located between the asci. Locate asci with their characteristic eight ascospores.

Powdery mildew

Obtain some leaves or fruit that have powdery mildew on them. Place a drop of water on the infected area, and scrape off some of the powdery material and hyphae into the drop of water. Place the material on a slide, and observe the asexual spores with low and high power.

If a dry leaf with powdery mildew is available, repeat the process on

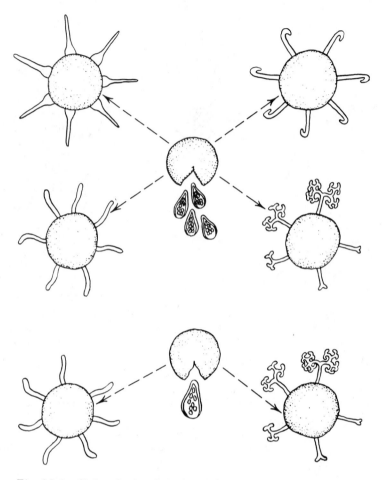

Fig. 25-4. Cleistothecia of six species of powdery mildews. *(Reproduced with permission from Weier, Stocking, and Tucker,* Botany, a Laboratory Manual, *1982. Courtesy John Wiley and Sons.)*

this leaf. In your scraping you will have removed a cleistothecium (Figure 25-4), a type of completely enclosed ascocarp with appendages.

After observing the ascocarp under low power, firmly press down on the cover slip. This will rupture the ascocarp and release the asci. Determine the number of ascospores in each ascus. The number of spores in the ascus is specific for each genus.

Penicillium

One of the species of *Penicillium* is the blue or green mold that grows on oranges. Prepare a wet mount of some of the hyphae from the edge of the growth. Ascomycota reproduce asexually by means of special spores called **conidia**. Observe the arrangement of conidia either from the living material or in prepared slides. Conidia are not produced inside a sporangium, but at the tips of special hyphae.

BASIDIOMYCOTA

The Basidiomycota, the "club fungi," include some of the largest and most conspicuous fungi as well as some of the most poisonous fungi. This group includes puffballs, mushrooms, bracket fungi, and toadstools. Although the

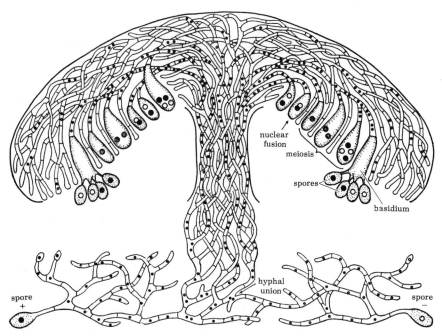

Fig. 25-5. Diagram of section through a mushroom. The entire stalk and cap are composed of hyphae packed tightly together. Spores are produced by basidia on the lower surface of the cap. *(From L. W. Sharp, Fundamentals of Cytology, McGraw-Hill Book Co., 1943. Used by permission.)*

aboveground portion of these organisms, the **fruiting body,** appears to be a solid mass of tissue, it is still composed of hyphae. The fruiting body represents a small portion of the entire organism—only the reproductive structure. Most of the organism is beneath the ground or growing through whatever substrate the fungus is using.

The hyphae in fungi of this phylum have perforated septa. Sexual reproduction is initiated by contact between compatible mating strains. The hyphae fuse, and a new generation of binucleate cells, formed in the region of the fusion, eventually grows into the large reproductive structure that may produce billions of spores (Figure 25-5).

Basidiomycota are characterized by their club-shaped reproductive cells, the **basidia.** Figure 25-5 shows the relationship of the basidium to the fruiting body as a whole, and Figure 25-6 shows several single basidia. The

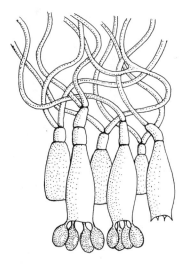

Fig. 25-6. Reproductive structures of Basidiomycota. The large club-shaped cells at the bottom are basidia. A mature basidium bears four round spores at its end. The three shorter basidia in the figure are immature. The one on the far right is an old basidium that has shed its spores.

cells of the hyphae that produce basidia are binucleate, containing one haploid nucleus from each of the mating hyphae. Terminal cells of these hyphae become the zygotes when their nuclei fuse in fertilization. The zygote develops into a basidium, and its diploid nucleus divides meiotically, producing four haploid nuclei. Each nucleus migrates into one of the four small protuberances developed at the end of the basidium. The tip of the protuberance then becomes walled off as a **basidiospore**, which falls off or is ejected. Each basidiospore can germinate into a new hypha.

Examine on demonstration the fruiting bodies of various members of Basidiomycota.

With your dissecting microscope, examine an edible mushroom. Note the gills, which are located under the umbrella-shaped cap. Basidia line the gills.

Examine a cross section through the cap of a mature *Coprinus.* The small dark structures on the edges of the gills are the basidiospores.

QUESTIONS FOR FURTHER THOUGHT

1. Although the spore stages of the fungi are emphasized, do the spores really damage the plants? If not, what is the function of the spores, and what does the damage?

2. What would be the advantages and disadvantages if fungi produced only asexual spores?

3. What are the main differences among the four fungal divisions?

4. Compare *Rhizopus* and *Spirogyra* in regard to their habitat, structure, and sexual reproduction.

5. What gives bread mold its black color?

TOPIC 26 CNIDARIA (COELENTERATA)

Biological Science, 4th ed.
pp. 1105–1108.
Elements of Biological Science, 3rd ed.
pp. 676–680.

Coelenterates include forms such as hydrozoans, jellyfish, sea anemones, corals, and their relatives; almost all of them are marine. The life cycle of some coelenterates includes a **medusa** stage and a **polyp** stage. The medusa has the form of a jellyfish (Figure 26-1), is free-floating in the water, and usually produces gametes. Polyps are fleshy columns with an attachment at one end and a mouth and tentacles at the other. Medusas are often produced by a polyp specialized for this purpose.

The members of the phylum Cnidaria are radially symmetrical with a body wall consisting of two prominent cell layers: an outer **epidermis** and an inner **gastrodermis**. Between the two primary cell layers is the mesoglea (mesoderm). The epidermis contains **gland cells, nematocysts,** and **epitheliomuscular cells**. Fibers of the epitheliomuscular cells form the basis for movement in coelenterates since there are no specialized muscle cells.

Food is taken in and waste eliminated through the same opening. This opening leads to the **gastrovascular cavity**, so-called because both partial digestion and circulation occur there. No special circulatory system is present, but cilia on the gastrodermal cells cause food in the cavity to move about, thereby reaching all parts of the organism, including the interior of the tentacles. Digestion is partly extracellular; the gastrodermal cells phagocytize and digest intracellularly the particles resulting from extracellular digestion.

The nervous system is in the form of a net, which is unpolarized; impulses can travel in either direction along the nerve cells. The nerve net is modified slightly in some coelenterates such as jellyfish and sea anemones, which are capable of more complex coordinated movement.

In this laboratory, we shall examine one of the few freshwater species, hydra, and several other members of this phylum to illustrate the general body plan, cellular construction, and behavior.

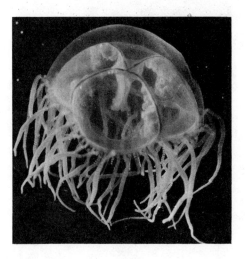

Fig. 26-1. Jellyfish. *(Courtesy Carolina Biological Supply Company.)*

Fig. 26-2. Hydra, with bud. *(Courtesy Ward's Natural Science Establishment, Inc., Rochester, New York.)*

STRUCTURE OF THE HYDRA

The hydra is a small coelenterate that lives attached to submerged rocks, leaves, and twigs, its tentacles outstretched for passing prey. Obtain a specimen in a small petri dish, or place one on a slide with a well-raised cover slip. Allow the animal to relax, and examine it under the dissecting microscope and, using low power, under the compound microscope. Note the radial symmetry. Locate on your specimen and label in Figure 26-2 the following:

Examining a hydra

1. Basal disc (the animal's point of attachment).
2. Body column.
3. Tentacles.
4. Mouth region (at the base of the tentacles).
5. Nematocysts (small swellings localized largely in the tentacles).

Locate and label in Figure 26-3 the epidermis and gastrodermis. If your specimen is mounted on a slide, you may be able to see the epidermis

Fig. 26-3. Hydra, cross section. *(Courtesy George H. Conant, Triarch Incorporated, Ripon, Wisconsin.)*

and gastrodermis with the compound microscope. In the hydra and in most other members of its class (the Hydrozoa), there is a small amount of jelly-like material, the mesoglea (mesoderm), between the epidermis and gastrodermis. In many other coelenterates (e.g., sea anemones), the mesodermal layer forms most of the animal's bulk.

In the hydra, there is no marked nervous centralization, although there is a tendency for nerve cells to congregate about the mouth region. You can determine that the hydra is more sensitive around the mouth by touching the animal gently with a teasing needle in several regions, including the mouth.

FEEDING BEHAVIOR IN THE HYDRA

Prey capture   All coelenterates are carnivorous predators which capture live prey by means of nematocysts, specialized epidermal cells on the tentacles. When properly stimulated, nematocysts discharge threads that entangle or pierce and poison the prey. Following capture, the victim is ingested and digested; any indigestible remains are eliminated through the mouth.

1. Transfer your hydra to a small petri dish or a Syracuse dish, and observe the animal under the dissecting microscope, preferably with a dark background and side lighting. Introduce some small brine shrimps (artemia) that have been carefully rinsed free of sea water; watch the hydra carefully for the next few minutes. Record your observations in as much detail as possible. (*Note:* If the hydra has not captured and ingested one of the brine shrimps within a few minutes, return your animal to the instructor and obtain a fresh specimen.)

2. If additional specimens are available, you may wish to compare your results with the feeding responses of hydras that have already been fed.

Feeding response   1. The capture of prey is normally followed by a relatively complex series of movements in which the prey is transported to the widening mouth and then pushed into the gastrovascular cavity. If your specimen did not respond in this way, watch another unfed animal capture and ingest prey. These actions constitute the "feeding response" and are apparently chemically controlled.

At present, it is unclear whether the chemical message arrives from the prey or is a result of nematocyst discharge or both. The feeding response can be induced by a variety of chemical agents, some of which are present in the body fluids of prey animals. A coelenterate would be expeosed to these substances when its nematocysts pierce the prey. But experiments have shown that an animal that has adapted to one of these chemical agents and is no longer showing the feeding response can be induced to respond when it is allowed to capture normal prey.

2. Your instructor will provide a solution of reduced glutathione, an agent that not only induces the feeding response in a hydra, but is also present in the body fluids of prey animals. Set up a properly controlled experiment to test the hydra's reaction to the presence of reduced glutathione.

How you use the solution in the experiment is up to you. But be sure not to add more than a few drops, and be careful to include appropriate controls. Record the results.

3. Using a new sterile lancet, obtain a large drop of blood from your finger, and let the blood dry on a slide. Discard the lancet immediately after use; do *not* leave it lying on the laboratory desk. After the clot has dried, break it up with forceps, and offer one of the larger pieces to the hydra. When the animal has "captured" the blood clot, let the clot go. Observe the reaction.

FEEDING BEHAVIOR IN A SEA ANEMONE

*Observing a
sea anemone feeding*

Obtain a starved sea anemone (such as *Metridium* or *Anthopleura*), placing it in a large culture dish of sea water.

Your instructor will provide clam juice and bits of clam meat, sand, glass rods and medicine droppers, and filter paper (which may be soaked in meat juice, sea water, saliva, or other substances). What is the reaction of the tentacles to the various forms of mechanical and chemical stimulation? What stimuli are necessary for a feeding response? How does the anemone compare in this respect to the hydra? What events constitute the feeding response in the anemone?

NEMATOCYST DISCHARGE

Hydra

Transfer a hydra to a slide containing a small amount of water. Add a cover slip to the slide, and put the slide aside to dry slightly. Observe the hydra at intervals under the low and high power of the compound microscope. The appearance of discharged and undischarged nematocysts is shown in Figure 26-4. Find as many types as you can on your slide, and make sketches of them.

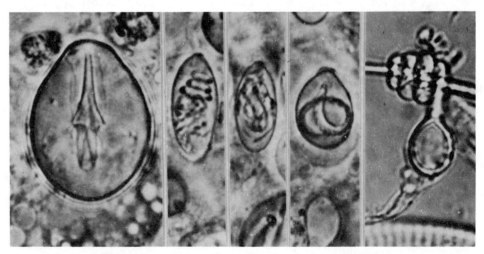

Fig. 26-4. Nematocysts. The one on the far left penetrates prey, while the others entangle prey with sticky threads. The one on the far right has been ejected and is shown attached to the bristle of a prey animal. *(Courtesy Carolina Biological Supply Company.)*

Sea anemone Using several simple experiments, best carried out with a sea anemone, we can demonstrate the nature of the stimuli responsible for nematocyst discharge.

1. *Mechanical stimulation.* Isolate a tentacle from an anemone by using forceps to grasp the basal portion and scissors to remove the tentacle completely. Carry out this procedure as quickly and gently as possible in order to avoid nematocyst discharge due to excessive mechanical stimulation. Place the tentacle in sea water in a depression slide, and examine the tentacle under low power. Stimulate the tentacle by gently rubbing it with a fine glass rod. Examine the tentacle for nematocyst discharge.

2. *Chemical stimulation.* With as little disturbance as possible, introduce to the above slide a drop of fresh clam extract or a bit of saliva dissolved in sea water. Again, examine the tentacle for evidence of nematocyst discharge.

Under normal circumstances, neither chemical nor mechanical stimulation alone is sufficient to cause nematocyst discharge. Combined, however, they will cause nematocysts to fire; you can demonstrate this by mechanically stimulating the tentacle in the presence of a chemical stimulant. Other factors, such as the nutritional state of the animal, also influence nematocyst discharge.

How would use of both chemical and mechanical sensory modalities, rather than just one or the other, be of advantage to the animal?

COELENTERATE DIVERSITY

Although the cellular structure and feeding behavior of the hydra and sea anemone are typical of coelenterates, the life cycle and many other features are not. Both of these animals are polyps. Many coelenterate polyps live in colonies. Each colony contains individuals specialized for several different functions, a condition known as **polymorphism**. Hydras and sea anemones are both specialized in their lack of the medusa stage; both produce gametes in the polyp stage.

Obelia 1. Both polymorphism and the medusa stage are illustrated by *Obelia,* a common marine form. Examine a slide of *Obelia,* and locate **feeding** and **reproductive polyps.** The feeding polyps are much like small hydras, with mouth tentacles and gastrovascular cavity; the reproductive polyps are covered with buds of medusas, which are produced asexually. The various polyps in the colony are interconnected by a hollow stalk. All, therefore, communicate with the gastrovascular cavity of the feeding polyps. Both stalk and individuals are largely covered by a hard material, the **perisarc.** In some species, the soft parts of the body can be withdrawn completely into the perisarc and thus protected from predators. Label Figure 26-5.

Gametes are usually produced by the medusa, except in forms like hydras or sea anemones, which reproduce sexually in the polyp stage. In either case, gametes are usually shed directly into the ocean. Early development produces a small **planula** larva, which eventually settles and develops into a new polyp.

2. Examine demonstrations of other members of three coelenterate classes:

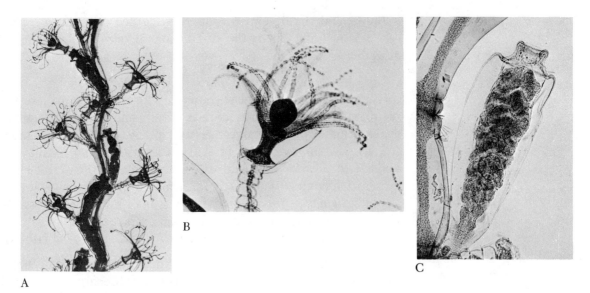

Fig. 26-5. *Obelia.* A: Part of a mature colony; B: a feeding polyp; C: a reproductive polyp. *(A: Courtesy George H. Conant, Triarch Incorporated, Ripon, Wisconsin; B, C: Courtesy Ward's Natural Science Establishment, Inc., Rochester, New York.)*

a. Hydrozoa.
b. Anthozoa.
c. Scyphozoa.

Review and list the characteristics of each class, and enter in the appropriate section of Appendix 2 each of the animals observed.

QUESTIONS FOR FURTHER THOUGHT

1. Distinguish between polyps and medusas. What is the function of the medusa in the coelenterate life cycle?

2. What role might diffusion play in coelenterate metabolism?

3. What is polymorphism? How might it benefit a coelenterate?

4. What is the mesoglea?

5. Summarize your findings in the experiments on feeding behavior in the hydra and sea anemone, and suggest experiments to answer some of the questions raised by your studies.

6. Marine organisms that live near the shore are often less sensitive to small environmental changes than are freshwater organisms or marine organisms that live far from shore. What adaptive value would a high tolerance to environmental change have for those organisms that live close to the shore?

Biological Science, 4th ed.
pp. 1112–1116.
Elements of Biological Science, 3rd ed.
pp. 680–688.

TOPIC 27 PLATYHELMINTHES

Members of the Platyhelminthes, a phylum of dorsoventrally flattened worms, occur in both marine and freshwater habitats, and several have adaptations for living in moist places on land. Some of the marine species are conspicuous and colorful; most of the freshwater forms are small, retiring, and easily overlooked in the wild.

While their bodies are constructed on a relatively simple plan, flatworms differ from Cnidaria (coelenterates) in that their mesoderm is a more prominent tissue layer. It forms muscle cells and associates with the other tissue layers to form distinct organs. Since flatworms have no cavity or coelom between the digestive tract and body wall, they are called **acoelomates.**

There are three classes of flatworms. The free-living species comprise the class Turbellaria, whose feeding habits range from predation through scavenging of detritus. The other two classes—class Trematoda (the flukes) and class Cestoda (the tapeworms)—are serious parasites of man and a wide variety of other animals. In this unit we shall examine a freshwater planarian worm to get a general idea of the structure and behavior of members of this phylum. Then we shall turn to specimens from the other two classes, which illustrate some of the special adaptations required by parasitic organisms.

CLASS TURBELLARIA (FREE-LIVING FLATWORMS)

External structure of a planarian

The common planarian *Dugesia* inhabits freshwater streams and ponds, scavenging upon small live or freshly dead animals. Obtain a specimen, and place it in a small petri or Syracuse dish of spring water. In subsequent handling of these animals, use only spring water; tap water contains chlorine and other chemicals that are toxic to these delicate animals. Observe the following:

1. Gliding locomotion, powered by cilia that cover the animal's ventral surface and directed by muscular movements of the body. If you use a dissecting microscope, you will probably notice a slimy mucous trail; mucus is secreted by glands around the margin of the ventral surface of the animal, and it provides traction for the cilia.

2. Distinct bilateral symmetry, in contrast to the radial symmetry of coelenterates. What is the advantage of bilateral symmetry?

3. Dorsoventral flattening, with the dorsal surface arched slightly (best seen in prepared cross sections).

4. The head, somewhat pointed in *Dugesia,* and bearing a pair of pigmented eyes and a pair of lateral mechanosensory and chemosensory pro-

jections—the auricles. Notice how the moving animal "tests" the environment ahead of it by oscillating the head region.

Internal structure of a planarian

To see the internal structure of *Dugesia,* mount a living animal in spring water between two glass slides held apart by small pieces of modeling clay (Figure 27-1). A little pressure on the top slide will immobilize the animal by compressing it; either dorsal or ventral side may then be studied under the compound microscope. Too much pressure will kill the animal.

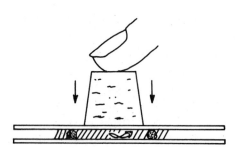

Fig. 27-1. Method of flattening a planarian worm. The worm is placed in spring water between two slides that are held apart by pieces of modeling clay. The slides are pushed together slowly by pressure on one of them until the worm is barely flattened and immobilized. Too much pressure will kill the animal.

1. The mouth is a small hole near the center of the ventral surface. It is sometimes difficult to see. The mouth leads into a short cavity containing the protrusible pharynx, which in turn communicates with the large gastrovascular cavity. Look for the pharynx by focusing just anterior to the mouth region.

The gastrovascular cavity branches into three main divisions. One runs anteriorly and terminates between the eyes, while the other two approximately parallel branches run posteriorly. All three branches subdivide repeatedly. To see the full extent of the gastrovascular cavity, examine a recently fed specimen or a slide made from such a specimen.

2. You may see a pair of ventral nerve cords by mounting and examining a living specimen of the white planarian *Phagocata,* or by examining a demonstration slide of *Bdelloura,* a marine flatworm. In the latter specimen the nerve cords show up clearly as a pair of longitudinal, often orange-colored structures terminating anteriorly as the brain. Flatworms show a distinct tendency toward cephalization or concentration of nervous tissue at the anterior end of the animal. This trend is carried further by the more complex phyla.

3. Flatworm gas exchange and excretion occur largely by simple diffusion. Water-balance problems are solved by a flame-cell system.

4. Reproduction is both sexual and asexual in most flatworms. Sexual reproduction is seasonal, and the reproductive organs develop in each individual only as the season approaches. Male and female organs are present in each worm, but the animals mate with one another and cross-fertilization, through exchange of sperms, is thus assured.

Behavior of a planarian

1. *Feeding.* Flatworms are notoriously sensitive to disturbances in the water about them, but you may be able to induce your animal to feed. When food is encountered, the pharynx is protruded through the mouth, and slight sucking action draws food particles into the gastrovascular cavity.

Place a small piece of hard-boiled egg yolk into a dish containing one or two specimens. Position the dish under a dissecting microscope, and shield the dish from light and from other disturbances. Then go on to another part of this unit, inspecting the dish at intervals to see if the worms are feeding. When you see the worms feeding, look for the protruded pharynx. If you cannot find it, lift one end of the worm gently with forceps, and try to observe the pharynx before it is withdrawn.

2. *Response to light.* You can test the flatworm's response to light by placing several specimens in a petri dish of pond water and offering them a choice of light or dark regions. Cover one-half of the dish top with black construction paper, new carbon paper, or a book, and arrange a bright light to illuminate, without overheating, the other half of the dish (Figure 27-2). Start the test by placing more of the animals in the illuminated half of the dish. Note their response after a few minutes, and repeat the test several times for consistency (i.e., reintroduce animals to the light, by turning the dish cover so that they are again exposed). Of what value would these responses be in nature?

Fig. 27-2. Light-response chamber for testing the response of planarians.

By carrying out the same experiment after cutting the animal into two pieces,* one piece with and one without eyes, you can test whether the eyes of a flatworm are the only means of light perception. Test both regions for a response to light. Be sure to try several different light intensities and to allow ample time for a response to occur. Record and explain all responses to each different condition.

3. *Response to touch.* Touch the animal on the auricles and on other parts of the body, recording the response in each case. Vary the nature of the stimulus (e.g., glass rod, teasing needle, water stream from medicine dropper) and the intensity of the stimulus. Design and carry out an experiment to discover whether integrating activity of the brain is necessary to elicit a response.

4. *Righting reaction.* Turn an animal over with a stiff paper card, and carefully observe its response. How do you think it senses orientation with respect to gravity? If you cannot answer this question, find another student whose worm is swimming upside down along the water surface. If you have such a worm, point it out to your instructor and to your neighbors. The worm is able to swim upside down on the water surface because of surface

*Place the flatworm on a stiff white card, and cut the animal with a razor blade.

tension. By adding a small amount of detergent like Photoflo or Tween 20 to the water, you can reduce the surface tension and cause the worm to drop to the bottom of the dish. Does the upside-down swimming give you any hint about how the planarian senses orientation with respect to gravity?

CLASS TREMATODA (FLUKES)

Structure of an Adult Fluke *Prosthogonimus,* a common parasite of the oviducts and intestines of domestic and wild waterfowl, is typical of its class. Study a slide of *Prosthogonimus* with reference to Figure 27-3, and examine live specimens, if available.

The animal is covered with a **cuticle,** which protects it from enzymes and toxic substances often present in the peculiar areas that many flukes inhabit (e.g., urinary bladder, intestines, etc.). Two prominent muscular **suckers** attach the animal firmly to its host. Why would strong attachment be necessary? The digestive system is Y-shaped, as in planarians. But the single branch of the Y is merely a short esophagus, opening within the oral sucker; the other branches of the digestive tract are simple, long, unbranched sacs that run along most of the animal's length. The nature of the digestive tract should make it easy for you to guess the animal's diet. Essentially all the other prominent organs are reproductive structures. Other endoparasitic flukes are structurally much like *Prosthogonimus.*

Fluke Reproduction Attached within its host, the adult worm produces hundreds of thousands of eggs, each enclosed in a resistant capsule. Within the capsule, a tiny **miracidium** larva develops. When the ciliated

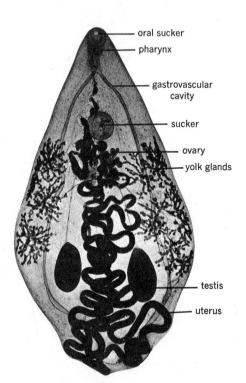

oral sucker

pharynx

gastrovascular cavity

sucker

ovary

yolk glands

testis

uterus

Fig. 27-3. *Prosthogonimus macroorchis,* a fluke that parasitizes the oviducts of the domestic hen and other waterfowl. *(Courtesy General Biological Supply House, Inc., Chicago.)*

miracidium hatches, it swims about until it locates an appropriate **intermediate host,** often a mollusc. There the miracidium undergoes a series of asexual multiplications, each one producing more infective units, which are eventually released from the mollusc and seek new intermediate hosts. The precise number of asexual stages and the precise hosts involved depend on the species of fluke.

Living flukes   If possible, you will be provided with a living frog, in which adult flukes may be found at several locations, especially in the lungs and in the bladder.

1. To find lung flukes, open the frog's body cavity, cut a lung at its base, and cut or tear the lung open. Pin the tissue to a wax-lined dissecting pan, and examine the tissue with a dissecting microscope. To locate bladder flukes, excise, open carefully, and pin the bladder (in a stretched condition) to a dissecting pan. Study its interior surface with a dissecting microscope.

2. When you locate a fluke, study its movements, and note how tenaciously it clings to its host. Examine the lining of the infected organ, noting the areas of hemorrhage and scarred tissue where flukes have previously been attached. What damage is done to the host?

3. Carefully remove an adult fluke, put it on a slide in a drop or two of pond water, and add a cover slip. Place the slide under a dissecting microscope. Then, with a piece of absorbent paper touching the edge of the cover slip, remove excess water until the worm is immobile and well flattened but not crushed. During the study, make sure that you add enough water to replace that lost through evaporation. Locate and identify as many of the internal structures of the fluke as possible.

4. Miracidia can be observed by teasing apart a mature bladder fluke in a small dish of pond water. The ripest eggs hatch almost at once upon contacting a hypoosmotic environment. Remove some miracidia to a slide for closer microscopic observation.

5. If materials are available, your instructor may provide other stages in the life cycle of the fluke.

CLASS CESTODA (TAPEWORMS)

Tapeworms are usually intestinal parasites of vertebrates. Their life cycle involves one or two intermediate hosts, and the tapeworms are highly modified for reproduction. Examine a living or preserved tapeworm. Note the attachment point or **scolex,** the short neck, and the long, ribbonlike body, consisting of increasingly mature segments called **proglottids.** New proglottids are produced at the base of the neck; as they grow older and farther from the scolex, they become sexually mature. The sexual organs can be identified in Figure 27-4.

Each sexually mature proglottid is inseminated by a more anterior one. It then begins to develop eggs. Ripe proglottids, which may contain more than 100,000 eggs, leave the host in feces. They hatch if eaten by an appropriate intermediate host. The larvae encyst in the host's tissues; the adult

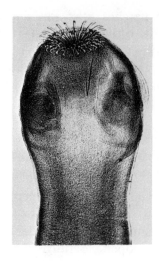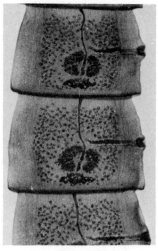

Fig. 27-4. *Taenia pisiformis,* the dog tapeworm. Left: Scolex (note the hooks and suckers); right: mature proglottids. *(Courtesy Carolina Biological Supply Company.)*

form develops when one of these cysts, present in raw or poorly cooked meat, is ingested by the appropriate final host.

Feeding behavior of tapeworms

How does a tapeworm feed? How do its feeding habits differ from those of flukes? How does damage to the host inflicted by a tapeworm differ from that done by flukes? Would the feeding habits of the tapeworm render it more vulnerable to elimination from the host than the feeding habits of flukes?

Enter the animals observed in this part of the laboratory in the appropriate spaces in Appendix 2.

QUESTIONS FOR FURTHER THOUGHT

1. What is the advantage of a flattened body?
2. Explain the term "cephalization" and how it applies to the flatworms.
3. How do flame cells function?
4. What adaptations allow some members of Platyhelminthes to maintain a parasitic way of life?

MOLLUSCA, ANNELIDA, AND ARTHROPODA

Biological Science, 4th ed.
pp. 1116–1119, 1122–1139.
Elements of Biological Science, 3rd ed.
pp. 684–685, 688–699.

In an earlier exercise in animal development, you learned that gastrulation is characterized by the formation of three germ layers—the ectoderm, mesoderm, and endoderm—and an archenteron, or primitive gut, which is formed by morphogenetic movements. The opening to the archenteron is the blastopore. In Cnidaria and Platyhelminthes, the blastopore becomes the opening into the gastrovascular cavity. In the phyla Mollusca, Annelida, and Arthropoda, the blastopore will develop into the mouth. Organisms that follow this path of development are referred to as the **Protostomia**. Animals such as echinoderms and chordates, in which the blastopore develops into the anus (and then the mouth develops later at the opposite end), are called **Deuterostomia**.

Protostomia and Deuterostomia also differ in the origin of the mesoderm layer and in the manner that the coelom forms. In the protostome phyla, the coelom forms from a split in the solid mass of mesodermal cells, whereas in the deuterostomes, the coelom develops in the mesodermal sacs, which evaginate from the wall of the archenteron.

In this laboratory, we shall examine the three major protostome phyla to help you understand the distinctive characteristics of each of these phyla and the diversity of animals they encompass.

MOLLUSCA

The phylum Mollusca is the second largest in the animal kingdom. Though the members of the various molluscan classes differ considerably in outward appearance, they all have fundamentally similar body plans. The soft body consists of three principal parts: (1) a large ventral muscular **foot**, which functions in locomotion; (2) a **visceral mass** above the foot, which contains the various internal organs; and (3) a heavy fold of tissue called the **mantle**, which covers the visceral mass and which in most species contains glands that secrete a shell.

Class Cephalopoda   The class Cephalopoda includes squids, octopuses, and nautili, which are the most highly developed of the molluscs. These animals are extremely well adapted for rapid movement and a predatory life style. We shall examine the Atlantic coast squid *Loligo pealii* as a representative of the class Cephalopoda. Although each organ system will be dissected and observed separately, at times organs from other systems will be pointed out because of their position in the body.

External anatomy

Squids have the ability to change their body color rapidly in response to environmental conditions. This is due to the presence of chromatophores, or pigment bodies, which are visible on the surface of the body. These chromatophores are attached to muscle fibers, the contraction and relaxation of which control the size of the chromatophores, resulting in the color changes.

Examine the modified foot, which is made up of eight **arms** around the mouth and two long **tentacles**. Each of the arms and tentacles has several rows of suckers. In the adult male, the suckers of the lower left arm are modified for the transfer of sperms to the female.

When you hold back the arms and tentacles, you will see that a muscular membrane extends from the bases of these appendages toward the mouth. This membrane is divided into an outer part—the buccal membrane—and an inner peristomial membrane. The buccal membrane has muscular projections, each with suckers on the inner surface. The peristomial membrane surrounds the mouth opening through which the chitinous jaws can be seen.

Note the well-developed **eyes** located on the head. Identify the lens, iris, pupil, and cornea. At the ventral edge of the eye is a tiny opening called the aquiferous pore, which is used for equalizing pressure around the eye chamber. The olfactory crest, a folded and thickened integument, is dorsal to the eye. The olfactory groove is located under the olfactory crest (Figure 28-1).

The tough, muscular structure that encloses the remaining part of the body is the **mantle**. The open edge of the mantle is called the collar.

The flattened, cone-shaped structure extending from under the mantle is the **siphon**, or funnel. The siphon is a modification of the "foot" of this mollusc. Squids move rapidly, mainly using the mantle and the siphon. Water is drawn into the mantle cavity by contraction of the longitudinal muscles. Contraction of the circular muscles makes the collar region constrict closer to the head and forces water from the mantle cavity out through the siphon. The force of the water pushed out of the siphon propels the squid. The direction in which the siphon is pointing when the water is expelled partly determines the direction in which the animal will move.

Internal anatomy

Note the posterior ridge (Figure 28-1), which runs the full length of the mantle. Cut the mantle open along this ridge from the collar to the end of the cone, being careful not to cut the **visceral mass**, which occupies most of the mantle cavity. Pin the mantle to a dissecting pan. If the specimen is male, the organs should look like those illustrated in Figure 28-2. If you have a female specimen, a pair of nidamental glands, which secrete coverings for the egg capsules, will cover the rectum, ink sac, and part of the heart region. Remove these glands from the viscera, being careful not to pierce the ink sac.

In Figure 28-2, locate the **ink sac**. The ink sac, which is filled with melanin pigment, is attached by a membrane to the underside of the rectum in the ventral region, just below the siphon. Using a scalpel, remove the ink sac carefully. Place the ink sac in a petri dish containing water, and puncture the sac. The pigment is released by the squid when the mollusc is attacked and wants to escape a predator. The ink is used as a defense mechanism to inactivate for a short time the olfactory cells of predatory fish while the squid escapes.

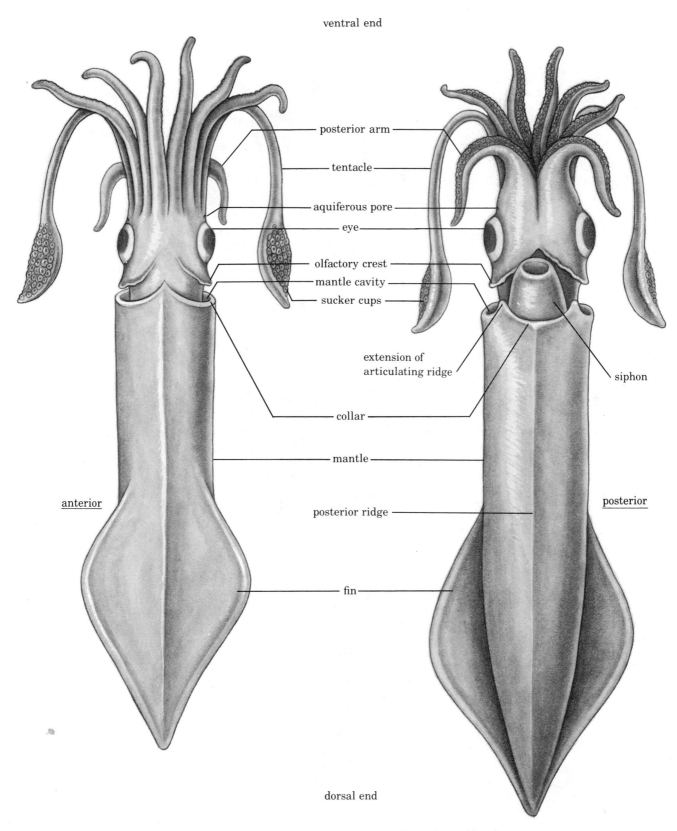

ventral end

posterior arm

tentacle

aquiferous pore

eye

olfactory crest

mantle cavity

sucker cups

extension of
articulating ridge

siphon

collar

anterior

posterior

mantle

posterior ridge

fin

dorsal end

Fig. 28-1. Anterior and posterior views of a squid.

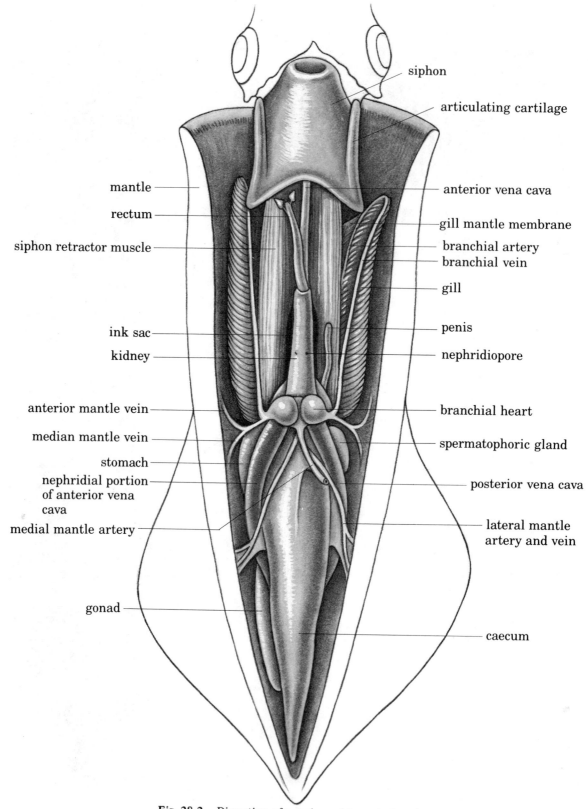

siphon

articulating cartilage

mantle

anterior vena cava

rectum

gill mantle membrane

siphon retractor muscle

branchial artery
branchial vein

gill

ink sac

penis

kidney

nephridiopore

anterior mantle vein

branchial heart

median mantle vein

spermatophoric gland

stomach

nephridial portion
of anterior vena
cava

posterior vena cava

medial mantle artery

lateral mantle
artery and vein

gonad

caecum

Fig. 28-2. Dissection of a male squid—posterior view.

Respiratory and circulatory systems

The **gills** are the most prominent structures in the mantle cavity. Oxygenated blood is transported from the gills through the **branchial veins**, large blood vessels along the outer edge of each gill, to the **systemic heart**. At this point it may be difficult to see the systemic heart since it lies under the kidney. From the systemic heart, the blood is pumped through the **anterior aorta** to the head region and through the short **posterior aorta** to the mantle arteries.

Venous blood from the head region is transported in the **anterior vena cava**. This blood vessel divides in the kidney region, at which point it is referred to as the nephridial portion of the anterior vena cava. Blood returns from the posterior regions of the body through the large, cone-shaped **posterior vena cavae**. The venous blood then enters the **branchial hearts** and is pumped to the gills to be oxygenated. Trace as much of the circulatory system as you can without removing any organs.

Digestive system

Cut off the siphon. Make a median longitudinal cut through the head at the base of the arms, and press the arms open. The hard, muscular, bulblike organ is the **buccal bulb** (pharynx). Remove the membranes around the mouth opening, and note the dark, chitinous jaws which resemble an inverted parrot beak. The squid uses these jaws to tear apart its food before digesting it. Using a sharp scalpel, remove the lower jaw, which will expose the **ligula**, a tonguelike structure. With forceps, pull the ligula back in order to see the **radula**, a food-rasping organ enclosed by the upper jaw. Rub a scalpel gently over the radula to feel the rough surface created by the numerous rudimentary teeth.

Follow the **esophagus** from the buccal bulb as it passes through the **liver** and enters the **stomach**. At the junction of the esophagus and stomach, locate the **pancreas**, a convoluted mass of tissue (Figure 28-3). Extending from the stomach is a membranous sac, the **caecum**. If the caecum is filled with digested material, it will expand to fill most of the mantle cavity.

On Figure 28-3 locate the **anal opening** which leads to the **rectum**. The **intestine** connects the rectum to the caecum. On the surface of the caecum near the connection to the intestine, there is a fan-shaped projection, the **spiral valve**. This structure separates food particles in the digestive system.

Nervous system

The ganglia, which look like white, pulpy masses, are situated in the head region on each side of the esophagus at about the level of the eyes. Outside the head region, on the surface of the mantle, locate the large stellate ganglia near the end of each of the gills.

The eye of the squid is similar to the human eye, but it is not homologous. Remove the eye, and identify the optic nerve, which connects the eye to the optic ganglion.

Female reproductive system

If you have a female squid, the **nidamental glands** were removed earlier so that you could observe the other organs. The **ovary**, when filled with eggs, is the largest organ and lies in the coelomic cavity. The eggs pass from the ovary into the mantle cavity and then to the large **oviduct**, which is parallel to the rectum. In the region of the branchial hearts, the oviduct is modified into an **oviducal gland**, which encapsulates the eggs. Eggs are released as egg clusters and are encapsulated by nidamental-gland secretions. Fertilization may occur before or after encapsulation.

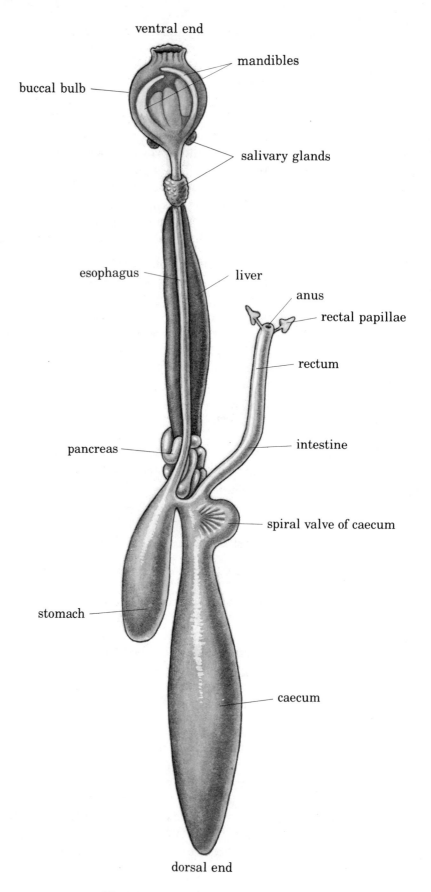

ventral end

mandibles

buccal bulb

salivary glands

esophagus

liver

anus

rectal papillae

rectum

pancreas

intestine

spiral valve of caecum

stomach

caecum

dorsal end

Fig. 28-3. Digestive system of a squid.

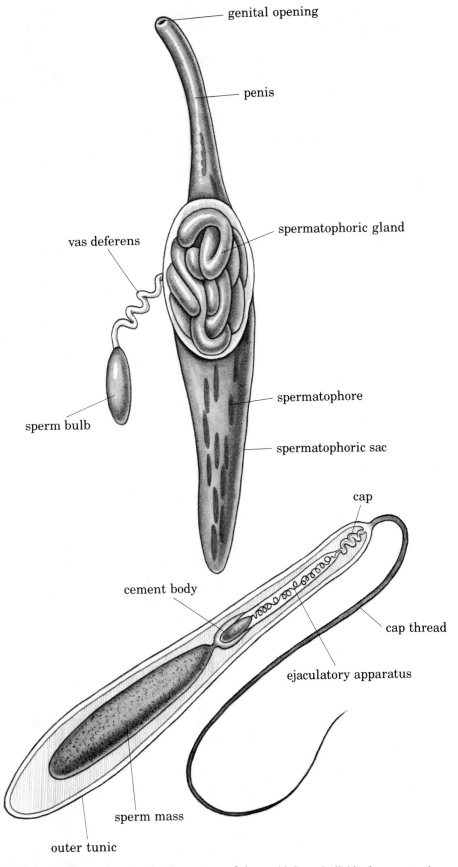

genital opening

penis

spermatophoric gland

vas deferens

spermatophore

spermatophoric sac

sperm bulb

cap

cement body

cap thread

ejaculatory apparatus

sperm mass

outer tunic

Fig. 28-4. A: The male reproductive system of the squid; B: an individual spermatophore.

Male reproductive system (Figure 28-4)

The **testis**, a light-colored, flattened organ, is located near the caecum in the coelom. To the right of the testis, near the branchial hearts, is the spindle-shaped **spermatophoric sac**. The coiled **vas deferens** is adjacent to the spermatophoric sac. The enlarged end of vas deferens is called the **sperm bulb**. The sperms enter the sperm bulb and then move into the vas deferens. The vas deferens connects to an enlarged convoluted tubule—the **spermatophoric gland**—which provides secretions and packages the sperms into **spermatophores**. The spermatophores move from the spermatophoric gland through a duct to the spermatophoric sac (seminal vesicle). The spermatophores leave the spermatophoric sac through the **penis**. The male transfers the spermatophores by a modified arm to the female.

With dissecting needles, remove some of the spermatophores from the spermatophoric sac, and make a wet mount of the spermatophores. Observe the spermatophores with a compound microscope (Figure 28-4).

After you have finished examining the internal organs, remove them, and dissect out the internal shell—the **pen**—from the mantle.

Examine members of the molluscan classes Amphineura (i.e., Caudofoveata, Solenogastres, Polyplacophora), Gastropoda, and Scaphopoda on demonstration, and read the explanatory material beside each display. Compare and contrast the classes with regard to shell characteristics, method of feeding, and method of locomotion.

ANNELIDA

The phylum Annelida, or segmented worms, is divided into three classes: Polychaeta, Oligochaeta, and Hirudinoidea. We shall look at some of the characteristics that distinguish the classes.

Class Polychaeta   Examine the polychaetes on demonstration. Can you see a well-defined head? What sorts of sensory structures does it bear? The body segments of these marine worms usually bear lateral appendages called parapodia. What is their function? Some biologists have suggested that arthropod legs evolved from annelid parapodia.

Class Oligochaeta   Examine the oligochaetes on demonstration. This is the class to which earthworms belong. The class also contains many freshwater species.

The body of the earthworm is composed of a series of distinct segments, each of which is partitioned internally from adjoining segments by a membrane.

Examining an earthworm

1. Obtain a live earthworm, and identify the dorsal side by the presence of a dark, pulsating line—the **dorsal blood vessel**. Insensitize the animal by placing it in 35% alcohol for a few minutes; the worm will writhe vigorously at first, but after a few minutes the writhing will cease and the worm will respond only slightly to gentle probing. Remove the worm from the alcohol

as soon as it no longer responds to the probing (this will take several minutes). If left in the alcohol for a longer period, the worm may expire.

2. Pin the worm *dorsal* side up through the third or fourth segment and again through the body about 3 inches behind the anterior end. Fasten the specimen to a dissecting pan in such a way that the worm will be observable when the pan is later placed under a dissecting microscope. (If you pin the worm in the middle of the pan, the specimen often will not be visible under the dissecting microscope.)

3. With fine scissors, make an incision in the dorsal mid-line of the worm, beginning just in front of the **clitellum** (Figure 28-5) and proceeding anteriorly. Cutting tiny blood vessels in the body wall is unavoidable, but keep the scissors-point up to avoid damage to fragile underlying structures.

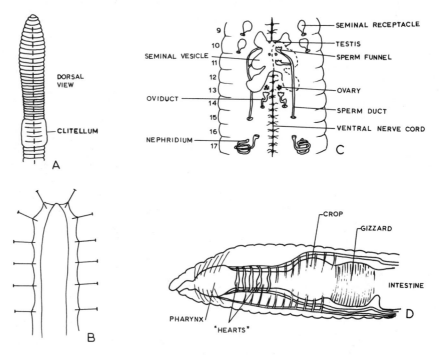

Fig. 28-5. Earthworm. A: Outline of incision; B: method of pinning the worm to the pan; C: reproductive system and parts of the nervous and excretory systems, all in dorsal view with segments numbered; D: anterior parts of the circulatory and digestive systems in lateral view.

4. When the worm is opened dorsally up to the anterior pin, spread the body wall, breaking the septa between segments, and pin the body wall to the dissecting pan as illustrated in Figure 28-5. Cover the dissection with Ringer's solution.

5. Locate the dorsal blood vessel, partly embedded in the top of the intestine, which runs most of the length of the worm. The dorsal blood vessel is the main pumping device and is supplied with valves, the action of which you can see by examining the vessel with a dissecting microscope. In which direction does blood move in the dorsal vessel?

6. In segments 7 through 11 (count from the anterior end), most of the blood in the dorsal vessel passes ventrally through five pairs of **aortic arches**

(or "hearts"), which supplement the action of the dorsal vessel in moving blood through the system.

7. Remember that blood travels through the worm without ever leaving the vessel system. In what regions of the worm's body would you expect to find the most well-developed capillary circulation? Examine tributaries of the dorsal and ventral blood vessels to see if your hypothesis is correct.

8. Review, with the aid of Figure 28-5, as many other structures in the earthworm as time permits.

The animals in several phyla with well-developed circulatory systems have evolved excretory organs known as **nephridia**. These remove wastes from both the tissue fluid and the blood, and thus exhibit some properties similar to flame cells and some properties similar to kidneys. Except for a few anterior segments, each compartment of the earthworm thus formed has its own pair of nephridia, which open independently to the outside.

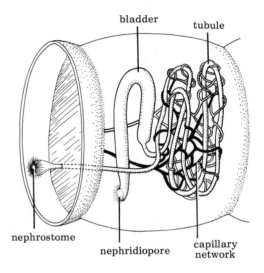

Fig. 28-6. Earthworm nephridium.

Referring to Figure 28-6, notice that the nephridium consists of several parts. Fluids first enter a ciliated funnel, the **nephrostome**. The beating of cilia moves the fluids into a long **tubule**. A great many capillaries are located around the tubule, and it is here that exchanges occur between the blood and the nephridium. Wastes are removed from the capillaries, and presumably there is some reabsorption of water and other substances from the tubule. The tubule eventually empties into a large distensible **bladder** with muscular walls. This storage structure, in turn, empties to the outside via a ventrally located pore—the **nephridiopore**.

Note that the nephrostome is located in the segment immediately anterior to the one containing the rest of the nephridium. Each nephridium thus picks up tissue fluid from one segment and collects wastes from the blood in the next posterior segment.

Choose a section near the middle of the worm, arrange the pins so that the membranes between segments are stretched as little as possible, and identify the above-mentioned parts of the nephridium. Careful dissection under the microscope will yield the whole nephridium (which can be wet-mounted on a slide and examined under the compound microscope) or its parts: the beating of flagella can be seen in the nephrostome, capillaries can be observed in association with the tubule, and parasitic roundworms can often be seen in the bladder.

When you have completed your dissection, unpin the worm and discard it; clean, rinse, and return the pins to the container at your desk; and rinse and dry the dissecting pan.

Class Hirudinoidea Most leeches are specialized as external parasites. What are their most obvious special adaptations for parasitism? How do these animals move? Can you see any evidence of external segmentation?

ARTHROPODA

Class Crustacea The arthropod class Insecta is by far the largest class in the entire animal kingdom. Its species inhabit almost every terrestrial habitat and many freshwater habitats as well. They are the arthropods most familiar to the average person. But since we have already examined insects in previous laboratories, we shall devote most of our attention here to the arthropod class Crustacea, another enormous group whose species also live in a great variety of habitats. Because most of those habitats are in the sea, crustaceans are not as well known to most people as insects are, but they are important in marine communities. Some representatives of Crustacea (e.g., crayfish) inhabit fresh water, and a very few (e.g., sow bugs) are terrestrial.

We shall examine in some detail a crayfish or a lobster. We must emphasize, however, that the characteristics of these animals are different from those of many other crustaceans, and you should not think of crayfish and lobster as "typical" members of their class. Indeed, the great majority of crustaceans are planktonic organisms that bear little resemblance to a lobster.

External anatomy Place a crayfish or a lobster in a dissecting pan. Note that the entire body of the animal is covered with a hard **exoskeleton**. The exoskeleton, which is composed of chitin (a polysaccharide) and clacium carbonate, is secreted by the underlying epidermis. What are some functions of the exoskeleton?

The body is divided into an anterior **cephalothorax** and a posterior **abdomen**; the exoskeletal shield covering the cephalothorax is called the **carapace**. In the adult animal, segmentation is evident only in the abdomen, but some other adult crustaceans and the early developmental stages of all arthropods show that segmentation is part of the basic body plan. What other phyla show marked segmentation? Is it always evident in the adult stages? What function does segmentation serve in arthropods? What function does segmentation serve in other animal phyla?

Each segment, including those fused to form the cephalothorax, bears an appendage. Though different in function, the appendages on successive segments develop embryonically in the same manner and are, therefore, spoken of as **serially homologous**. As each appendage develops, it becomes structurally different from the others; in the adult, each is modified to carry out some particular function.

1. Notice the five pairs of large appendages attached to the cephalothorax. The most anterior pair are **chelipeds** (pincers), which function in the capture of prey, in defense, or (in the male) for holding a female during mating. The posterior four pairs are called **walking legs**.

2. Behind the walking legs, on the abdomen, are five pairs of **swimmerets**. In the female, the swimmerets are much alike; in the male, the anterior two pairs are stiffened and folded forward, and aid in the transfer of sperms during mating. Pick up an animal of each sex to see these appendages clearly. (Press your index finger down on the carapace, and grasp the animal around the cephalothorax with the rest of your hand.)

3. The last abdominal segment bears a pair of broad, fan-shaped **uropods**. A terminal extension of the body, the telson, together with the uropods, forms a tail fan employed in rapid backward locomotion.

4. Anterior to the chelipeds are six pairs of serially homologous appendages, which are used mostly to handle food. With forceps, gently locate each appendage, and separate it from the others. The table below lists these appendages in order (from the cheliped forward):

APPENDAGE	FUNCTION
3rd maxilliped	
2nd maxilliped	
1st maxilliped	
2nd maxilla	
1st maxilla	
mandible	

Study each appendage carefully, and, in the right-hand column of the table, suggest its specific function. Later, you will have an opportunity to test these ideas. If you have time, you may later wish to remove these appendages and study their structure more carefully with a dissecting microscope.

5. In front of the mandibles is a final pair of homologous appendages—the antennae. On the basal segment of each antenna, locate a small raised nipple—the **excretory pore**.

The smaller **antennules** and the pair of stalked **compound eyes** are additional sensory organs that develop on the first segment. Early development shows that they are not homologous to the other appendages.

Between the bases of the third and fourth pairs of walking legs, the female has an exoskeletal swelling—the **seminal receptacle**—where sperms from the male are deposited during mating. At the bases of the second and fourth walking legs, respectively, locate openings of the female and male reproductive organs. The organs themselves may be seen in the internal dissection.

Behavior

Put a screen around a pan containing a living crayfish. Arrange the screen so that it will hide most of your movements. Dangle an earthworm near the animal's mouth parts, and describe in detail the crayfish's reactions, noting particularly the activity of appendages around the mouth.

Observe the locomotion of a crayfish, paying particular attention to the activity of the walking legs. Are all four legs identically constructed? Are all four actually used in walking? Do the swimmerets function in locomotion?

By making threatening movements or by pushing the animal gently from the front, you may induce it to leap backward with a sharp flexure of the abdomen. Of what value is this reaction?

Make threatening movements in front of several living crayfish until you see a "rearing up" response. To test their response to the same stimulation, try several other specimens of the same sex and different sexes. Is there any correlation between "aggressiveness" and the sex of the animal? If so, which sex is more aggressive and what function would this behavior serve in nature? What role does aggressive behavior play in societies where the male is the principal aggressor?

Turn your animal over in its pan, noting carefully its responses and the appendages it uses as it rights itself. Do any of the appendages seem to be consistently more active than others when the animal is turned upside down? Can you change this pattern of activity by removing or immobilizing the active appendages? What organ is responsible for sensing changes in the animal's orientation with respect to gravity?

Internal anatomy   Place a living crayfish in a small bowl of pond water. To handle a crayfish, immobilize it by pressing your index finger down on the carapace (Figure 28-7) and, with the adjacent pair of fingers, grasping the body just behind the claws. A crayfish held in this position is harmless.

Pick up the specimen. With fine scissors, make a cut in the carapace on one side, as shown by the dotted line in Figure 28-7. (This procedure is comparable to cutting fingernails.) Observe the feathery **gills**, attached at the base of each appendage in the thoracic region. Each gill branches into numerous filaments. Remove a small portion of a gill, and examine it under your compound microscope. Observe how the surface area is increased. How does gas exchange occur in the crayfish?

The carapace covers the gill chamber, protecting the fragile gills and isolating them from the environment. Thus, the gill chamber is open only at the anterior and posterior ends. Water is driven through it by the action of a

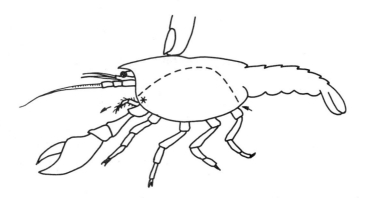

Fig. 28-7. Crayfish. Pressure from the index finger immobilizes the animal; the dotted line shows the cut to be made in the carapace; the asterisk indicates the position of the gill bailer; the arrows indicate the direction of the respiratory current.

part of the second maxilla, one of the mouth parts. We, therefore, call this appendage the **gill bailer**. It may be seen through the semitransparent carapace in the region just behind the obvious mouth parts. Its action can be noted in more detail if part of the carapace covering it is removed (this is most easily done on the side where a cut has already been made).

The current created by the gill bailer may be observed as follows:

1. Obtain a few dropperfuls of India ink in a beaker.

2. Draw up half a dropperful in a clean dropper, wipe the excess from the outside of the dropper, and, without touching the animal, release some of the ink underneath the posterior ventral portion of the carapace, as indicated by the rear arrow in Figure 28-7. (*Note:* If the animal moves away, gently hold it still with pressure on the carapace. However, the demonstration works best if the animal is unrestrained.)

3. Look for ink currents emerging anteriorly in the gill-bailer region.

Anesthetize your specimen by placing it in a fresh- or sea-water solution of chloroform for ten minutes or until the crayfish ceases to thrash about. Pin it securely, dorsal side up, in a wax-lined dissecting pan. Arrange the specimen so that it may later be viewed under the dissecting microscope without removing any pins.

With a teasing needle, scrape a small hole near the center of the dorsal side of the carapace. Alternately using needle, forceps, and fine scissors, carefully remove most of the dorsal carapace, as indicated by the broken line in Figure 28-8. Work forward and backward from your original hole, removing small pieces of exoskeleton as you go. Keep the instrument points up; tissues beneath the carapace are soft and easily injured. Once the animal has been opened, carry out under saline solution as much of the dissection as practicable, and make liberal use of your dissecting microscope.

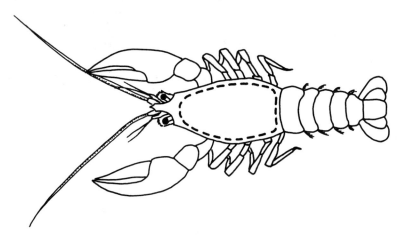

Fig. 28-8. Crayfish—dorsal view showing a cut in the carapace to expose the heart prior to internal dissection.

Circulatory system

1. You will now be observing the dorsal wall of the **pericardial sac**, which encloses the **heart**. If the animal was not overanesthetized, the heart will still be beating. If it is, carefully use forceps and scissors to remove the

pericardial sac. If not, inject a dye solution into the circulatory system, as described below.

Inject the heart with a concentrated dye solution, using a 1-cc dye-filled syringe and a fine-gauge needle. Insert the needle into the heart, but not all the way through it. Holding the syringe in place, rapidly depress the plunger. If the needle is properly in place, some dye will leave the heart through the organ's six surface openings (**ostia**), but most of the dye should fill the arteries. Remove the top of the pericardial sac (if it is not yet removed), and pour some water over the preparation to remove excess dye.

Remember that the crustacean circulatory system is an open one. Blood from the body drains into the pericardial sac and enters the boxlike heart through the three pairs of ostia during the heart's relaxation phase. The ligaments suspending the heart in the pericardial sac aid in expanding the heart's volume so that blood is pulled in during the relaxation phase. When the heart contracts, valves bounding the ostia prevent blood from leaving the heart and force the blood through several arteries. The ostial valves thus operate much like valves in the vertebrate heart to maintain a one-way circulation.

2. If your injection was successful, the ophthalmic and antennary arteries were almost certainly injected. The former artery runs middorsally from the anteriodorsal surface of the heart; the latter pair of arteries runs forward to serve antennae, anterior muscles, and other organs. Other arteries, which will be transparent unless injected, may be found posterior and ventral to the heart.

The arteries branch and rebranch, terminating in sinuses that surround the organs and in tissue spaces within the organs. These sinuses communicate with one another, and the blood eventually makes its way back to the pericardial sac.

Reproductive system

Remove the heart by cutting the arteries and ligaments that suspend it. Also remove the thin floor of the pricardial sac, just below the heart.

1. The reproductive organs of both sexes are approximately Y-shaped, with the upper branches of the Y extending forward and the lower branch running back toward the abdomen or into it. The Y shape will be clearest in the whitish testes. Near the center of the Y, a pair of sperm ducts arises, one going to the base of each of the last walking legs, where the duct opens to the exterior.

2. The ovaries are likely to be orange in color and filled with eggs. The size of the eggs and of the ovaries is determined by the season. Trace one of the two fragile oviducts to its opening on the base of the second walking leg.

3. If you remove the oviducts or sperm ducts and macerate them on a slide, you may be able to see living, mature sperms or eggs. If you find them, call them to the attention of other students.

Digestive system

Food enters the digestive tract through the mouth, an opening between the mandibles, and then passes through a short **esophagus** into a **cardiac stomach**, a **pyloric stomach**, and an **intestine**, which opens in the last segment of the abdomen as the **anus**.

1. The cardiac and pyloric stomachs are thin-walled, white structures near the dorsal mid-line of the cephalothorax. The former is the more an-

terior; by lifting up its anterior end, the esophagus is easily seen. The cardiac stomach bears in its interior a large number of chitinous plates, bristles, and teeth (together called the **gastric mill**) that serve to grind food. The grinding action is accomplished by the activity of many small muscles inserting on the wall of the stomach and originating on the exoskeleton. Locate some of these muscles as well as the large muscles that operate the mandibles; the latter will have been cut during the removal of the carapace earlier in the dissection.

2. Properly ground food is strained through a system of bristles into the pyloric stomach and then travels into the intestine or into one of the yellowish or yellow-green lobes of the digestive gland.

The digestive gland secretes enzymes into the cardiac and pyloric stomachs. Some digestion probably also occurs in the intestine and in the tubules of the digestive gland. In addition to providing enzymes, the digestive gland stores food. To what vertebrate organ would it be comparable? To what structure(s) in plants?

Trace the intestine posteriorly through the abdomen to the anus, paying attention to the blood vessels and muscles in the abdomen.

3. After the digestive system has been studied in place, remove it, and cut it open mid-ventrally under water. Study the arrangement of filtering and grinding plates and bristles. Postpone this removal, however, until after the nervous system has been studied.

Nervous system and sense organs

1. Carefully remove any remaining carapace, including the projection between the eyes; avoid injury to underlying tissues. Lift up the cardiac stomach, and, just ventral to a point between the eyestalks, locate a pair of large ganglia. These ganglia constitute the **brain.** If the circulatory injection was successful, the brain will be pinkish. From the brain, short nerve tracts run to the eyes, antennae, antennules, and dorsal surface of the head. These will show up clearly if a small amount of 95% alcohol is placed on them.

2. The largest pair of nerves leaving the brain runs around the esophagus to join the **subesophageal ganglion.** The rest of the central nervous system consists of a **ventral nerve cord,** a double structure that runs from the subesophageal ganglion back into the abdomen. To see the cord, remove all the overlying organs posterior to the esophagus. Also remove the calcified plates of the ventral cephalothoracic exoskeleton. These plates cover the nerve cord in the thorax, and they must be removed to expose it.

3. Antennae, antennules, and tactile bristles all over the body contain chemo- and mechanoreceptors. The **statocysts,** a single one of which opens on the basal segment of each antennule, help maintain equilibrium; each statocyst contains hair cells that are stimulated by the movement of tiny sand grains when the animal changes its position in space. How do these organs compare to your own equilibrium receptors?

4. Examine one of the compound eyes under the dissecting microscope, noting the square facets that make up the eye's surface. The separate units that comprise a compound eye are called **ommatidia**; you can see individual ommatidia if the eye is immersed in boiling water for a few seconds and then teased apart on a slide and examined microscopically. Compare your observations to Figure 28-9. How does the arrangement and operation of receptor

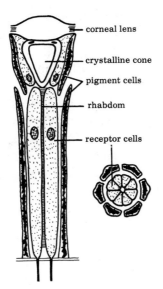

corneal lens

crystalline cone

pigment cells

rhabdom

receptor cells

Fig. 28-9. Longitudinal section and cross section of an ommatidium from an insect compound eye. The lens and crystalline cone focus incoming light rays into the rhabdom, which distributes the light into the eight receptor cells surrounding the rhabdom. The pigment cells prevent passage of light from one ommatidium to another. *(Redrawn from R. E. Snodgrass,* Principles of Insect Morphology, *McGraw-Hill Book Co., 1935. Used by permission.)*

cells in a compound eye differ from the arrangement of receptor cells and their operation in a vertebrate eye?

Excretory system   The excretory (and, in the crayfish, water regulatory) organs are a pair of large **green glands**, located just beneath the origin of the eyes. The soft, greenish portion is glandular and is overlaid by a transparent, membranous bladder. The bladder opens to the exterior through the excretory pore. With a fine-gauge needle and a syringe, inject a little dye or water into the bladder through the excretory pore to distend the bladder so that it will be more easily seen.

Other Crustacea   Examine other crustaceans on demonstration, and compare them with the crayfish or lobster.

OTHER ARTHROPOD CLASSES

On demonstration is a variety of representatives of other arthropod classes. As you look at them, ask yourself in what ways they resemble crustaceans and in what ways they differ. Also compare each class with the other classes on demonstration. Pay particular attention to body regions, antennae, mouth parts, other appendages, and special features such as wings, spinnerets, etc. There is explanatory material beside each demonstration.

Compare molluscs to annelids and arthropods with regard to gas exchange, circulation, excretion, nervous system, skeleton, and development.

QUESTIONS FOR FURTHER THOUGHT

1. What is the importance of plankton Crustacea in marine communities?

2. Discuss the assertion that the appendages of an arthropod tell you almost all you need to know about its habitat, what it eats, how it carries out gas exchange, how it moves, etc.

3. Describe the fundamental differences between a tapeworm, a flatworm, a roundworm, and an earthworm.

4. What is the adaptive value of a segmented body?

5. The larvae of certain clams attach themselves to the gills of fish. What adaptive advantage would there be to this arrangement?

6. In a sense, arthropods and vertebrates may both be regarded as the most highly evolved representatives of their respective evolutionary lines. They have independently evolved many similar "solutions" to problems of survival. Discuss the ways in which these two groups resemble each other and the ways in which they differ.

7. What do you think would have been the characteristics of the first arthropods?

TOPIC **29** ECHINODERMATA

Biological Science, 4th ed.
Chap. 42, pp. 1139–1146.
Elements of Biological Science, 3rd ed.
Chap. 33, pp. 699–705.

Deuterostomia includes four phyla: Echinodermata, Chaetognatha, Hemichordata, and Chordata. Echinodermata and Chaetognatha are invertebrate animals, Chordata consists of vertebrates, and Hemichordata has characteristics similar to both invertebrates and vertebrates.

Many aspects of the biochemistry, early development, and larval structure of echinoderms suggest that members of Echinodermata are more closely related to members of Chordata than to the other major invertebrate phyla, despite the fact that adult structure of echinoderms shows little resemblance to chordates. This does not mean that chordates evolved from echinoderms, but rather that the two groups diverged from a common ancestor.

Since the chordates have already been studied, we shall concentrate on the echinoderms in this laboratory. The sea star, as a representative echinoderm, will be examined in some detail, and examples of other echinoderm classes will also be on demonstration.

Members of the phylum Echinodermata are exclusively marine. Among them, besides the sea stars, are the brittle stars, sea urchins, sea cucumbers, sea lilies, and their relatives. Some echinoderms are used as food in tropical and Oriental cultures. Others (the sea stars) are important predators of commercially valuable molluscs such as clams and oysters.

CLASS STELLEROIDEA

Place a sea star in a deep dissecting pan, immersing the animal in sea water for as long as possible during the examination of external anatomy, and in cold tap water for the internal dissection. Using Figures 29-1 and 29-2, label the structures of the sea star with the appropriate names as you proceed with the following exercise in external anatomy.

External anatomy

Larval echinoderms, which you examined in Topic 21, are bilaterally symmetrical; the adults often show a five-part radial symmetry, after a complex metamorphosis. Your sea star will probably have five arms, or **rays**, attached to a **central disc**. This central region contains the main digestive tract and openings for the reproductive and water-vascular system. The rays contain the bulky digestive glands, reproductive organs, and apparatus concerned with the operation of the tube feet.

1. In the **oral** side of the central disc, find the mouth, protected by several pairs of spines. On the **aboral** side, opposite the mouth, find the prominent **madreporite**—a small, stony plate (often brightly colored) that

299

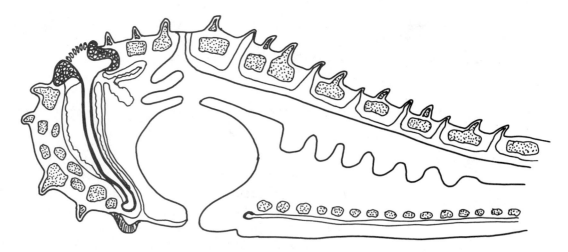

Fig. 29-1. Schematic longitudinal section through the central disc and ray of a sea star. Label all parts as you locate them when examining the external and internal anatomy.

Fig. 29-2. Schematic cross section through a sea-star ray. Label all parts as you locate them.

points away from you. The ray to the right of the madreporite is designated 1, and that to the left is designated 2; rays 3, 4, and 5 (and others, if present) follow in counterclockwise order. On the aboral side of the disc, between rays 4 and 5 on a five-rayed sea star, you may be able to locate the tiny anus.

2. With a dissecting microscope, examine the aboral surface of a ray and a disc. The animal is studded with calcareous spines. In Figure 29-1, note that these spines are projections from deeper-lying plates. The plates together form a protective endoskeleton, which originated from the embryonic mesoderm.

Close microscopic examination of the area around each spine will reveal a ring of **pedicellariae**—scissorlike devices. Stimulate a pedicellaria by drawing a thick thread across it or by poking it gently with a dissecting needle. What is the reaction? Of what value would these devices be to the animal in its natural habitat? Remove several pedicellariae to make a wet mount. Examine the pedicellariae with a compound microscope, and sketch what you see. Are all the pedicellariae the same? (*Note:* If sea urchins are available, examine their aboral surface to find the long, stalked pedicellariae and to test their reactions to various stimuli. In some urchin species, the spines and pedicellariae are associated with poison glands.)

3. In the spaces between the skeletal plates, note fingerlike sacs—surface projections of the body cavity. These are **dermal gills** which increase the surface area available for exchange of gases (and probably of water materials) between the body fluids and the environment. Cilia lining the spacious coelom (body cavity) keep fluids moving through the gills. And you may also notice movements of the gills themselves; your instructor may have injected your animal or a demonstration specimen with carmine dye particles. If so, look for red particles circulating within the gills; also remove the gill and examine it microscopically for evidence that the dye is picked up by gill tissues and excreted into the environment.

4. The surface of a sea star is fully ciliated. This can be verified by placing a small amount of carmine or carbon particles on the surface of a submerged animal. If you have time, you may wish to trace some of the patterns of ciliary movement on the rays and disc of a small sea star or to study similar patterns within the body cavity after you have begun dissection. The danger of colonization by plants and by sessile or parasitic animals is a real one for any marine organism, especially for one as slow-moving and defenseless as the sea star. Many marine animals and plants release gametes or larvae into the ocean; this ensures their dispersal but also attracts organisms seeking a place for attachment. External ciliary action is an important adaptation that prevents such settlers from making a home of the sea star. What other echinoderm adaptations would serve the same function?

5. On the oral side of each ray, locate the deep **ambulacral groove**. Longitudinal rows of **tube feet** (how many rows?) are located within it. The original function of tube feet in early echinoderms involved the procurement of food. Tube feet thus acted like tentacles, and in a few forms this function is retained. Most modern echinoderms, however, have developed the ability to use the tube feet for locomotion.

Each tube foot can be extended by the contraction of a balloonlike **ampulla** (Figure 29-2, not visible externally). A valve prevents water from reentering other parts of the water-vascular system. When the tip of the foot contacts a smooth surface, the foot adheres to the surface like a suction cup. Longitudinal muscles in the shaft of the foot then contract, shortening the foot and pulling the animal toward the point of attachment. The foot is then removed by contraction of radial muscles in its tip and reattached at another location. The action of a single tube foot does not have much effect on movement; but several hundred tube feet operating together in a coordinated manner serve very well for locomotion.

In sea stars, the tube feet are also used in feeding. The animal mounts a clam or an oyster and exerts a pull on the shells of the mollusc, thereby tiring the muscles that hold the shells together. Eventually, the shells open slightly, rendering the soft parts of the mollusc vulnerable to digestion by the sea star.

By means of a water-vascular system, water enters the tube feet through the madreporite and traverses a series of canals: the **stone canal**, leading orally directly away from the madreporite; the **ring canal**, running around the mouth; and **radial canals,** one entering each ray. Branches off the radial canals end in ampullae and tube feet.

6. At the tip of each ray, you will notice a pigmented spot that acts as a light receptor. Often the ray is upturned at the tip, exposing this region to the full light impinging on the animal.

7. The main parts of the nervous system (not visible externally) are a nerve ring and radial nerves, one passing ventrally down each ray between the tube feet. The responsibility of the nervous system for tube-feet coordination can be demonstrated by cutting the radial nerve in different locations on various rays and carefully observing the operation of the tube feet. If you have enough specimens and time for these experiments, turn in a full report of your findings.

Internal anatomy
(Figure 29-3)

Injured sea stars often cast off entire rays, and the missing rays are then regenerated. To ensure that your specimen does not cast off rays during the dissection that follows, inject each each ray (about halfway along its length on the aboral side) with 3 cc of 10% magnesium-chloride solution, which will serve as an anesthetic. Carry out your dissection under cold tap or sea water, and use the dissecting microscope whenever necessary.

1. Remove the aboral skeleton (the aboral surface of the sea star) from rays 3, 4, and 5, and from the central disc (avoiding the anus and madreporite) by cutting off the tips of the rays with stout scissors (do *not* use fine dissecting scissors except for internal structures) and cutting up both sides of each ray toward the central disc. Then, beginning near the tip of the ray, remove the aboral skeleton in small sections. Before you cut a piece of skeleton, lift

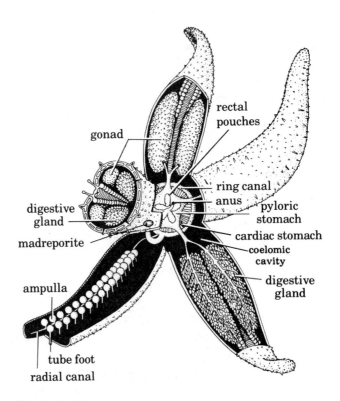

Fig. 29-3. Dissection of a sea star, showing major organs.

it up with forceps, and free any tissues that may be hanging in mesenteries from the skeleton. As you approach the central disc, join the cuts on adjacent rays, using stout scissors. Then remove most of the remaining skeleton of the rays and disc, again taking the skeleton off in small sections, being especially careful about freeing underlying tissues before removing the skeleton. In the central-disc region, you will have to cut the ducts leading from the gonads to the exterior (they empty via five pores) but leave the skeleton intact around the anus and madreporite.

2. The mouth leads into a short **esophagus** and then into a saclike, highly folded **cardiac stomach**. Above the cardiac stomach, the **pyloric stomach** receives five ducts, one from each **digestive gland**. Locate the digestive gland in each ray. Dorsal to the pyloric stomach, several brownish **rectal pouches** (probably excretory in function) can be seen. From the pyloric stomach, a short intestine leads to the small anus.

In most sea stars, the cardiac stomach is everted through the mouth during feeding and is applied to the tissues of the prey. Since the stomach is very membranous, it can be slipped with little difficulty into a small gap between the shells of a clam or an oyster. Proteolytic enzymes, secreted from the digestive glands, break down prey outside the sea star's body. Digestive products, along with bits of undigested food, are conveyed back into the sea star by ciliary action in the animal's digestive tract; some materials are also returned when the cardiac stomach retracts. A pair of prominent retractor muscles runs from the cardiac stomach into each ray. The cardiac stomach everts when these muscles relax and returns into the body when the muscles contract.

3. A pair of gonads (ovaries or testes) is located in each ray near the central disc and opens to the exterior via a pore. Testes are chalky-white or gray; ovaries are yellow or orange.

Remove a small amount of one of the gonads, and make a wet mount for microscopic observation. Both eggs and sperms are released into the ocean; fertilization and development occur outside the parent animals. Since the release of gametes must be reasonably simultaneous, sea stars (and the many other animals that reproduce in the same manner) have developed precise biological timing mechanisms. The stimuli involved are only beginning to be studied, but all are probably related to some natural cycle such as tides, the phases of the moon, etc. The nature of the receptors and the manner in which input becomes a signal for gamete release are poorly understood.

4. To see the **water-vascular system**, cut three or four of the digestive-gland ducts, freeing the stomachs from the rays. Cut the intestine (if it is not already cut), and push the stomachs gently out of the way so that the regions lying adjacent to them are clearly visible.

Find the stone canal, a calcareous tube leading orally from the madreporite. Lying along the stone canal is the axial sinus, a part of the sea star's greatly reduced circulatory system. The stone canal leads to a perioral ring canal, five radial canals, and then to the tube feet and ampullae.

Dispose of your specimen as directed by your instructor. Clean and dry the pan and syringe, and leave them neatly at your desk. Make sure that your dissecting instruments are well rinsed to rid them of salt water, and that they are carefully dried.

Examine other members of the class Stelleroidea.

OTHER ECHINODERMS

Representatives of other echinoderm classes are on demonstration. Examine them carefully, and read any explanatory material that may accompany the displays.

Class Crinoidea   Most sea lilies are sessile, being attached to the substratum by a long stalk. Unlike Stelleroidea and Echinoidea, the mouth is on the upper surface, where it is usually surrounded by long, feathery arms.

Class Echinoidea   Among the best known of this class are the sea urchins and sand dollars. They have no rays but do have five bands of tube feet. The plates of the endoskeleton are fused to form a rigid box or case, usually spherical, or flattened and oval. The body is covered with long spines.

Class Holothuroidea   These are the sea cucumbers. Their endoskeleton is much reduced, and the body often has a leathery texture. Unlike the other echinoderm classes discussed above, which rest on their oral surface, the sea cucumbers lie on their sides. The mouth is encircled by tentacles attached to the water-vascular system. Five prominent rows of tube feet run the length of the body.

QUESTIONS FOR FURTHER THOUGHT

1. Can you suggest some reasons why echinoderms are found only in the sea?
2. Where is the skeleton of an echinoderm? What function does the skeleton serve?
3. What characteristics distinguish the Deuterostomia from the Protostomia?
4. What method of reproduction is characteristic of echinoderms?
5. Some organisms, like nitrogen-fixing bacteria in the soil and green plants in both aquatic and terrestrial habitats, make obvious and important contributions to the economy of their habitats. What role do you think echinoderms play in the economy of the ocean?
6. How would radial symmetry be adaptive to a sea star?
7. Why does a sea star need a complicated system of channels for communication between the tube feet and the exterior? Why not just have a "sealed" water-vascular system?
8. Design the simplest possible experiment to measure the force exerted by a sea star to open a clam shell.
9. How do you think a sea lily feeds?

TOPIC 30 ADAPTATION

Biological Science, 4th ed.
pp. 864–881.
Elements of Biological Science, 3rd ed.
pp. 522–534.

From this course, two general facts should have been made especially clear. First, you know that all organisms face similar and interrelated problems in coping with their environments. They must have means of exchanging the raw materials and waste products of metabolism with their environment; they must transport such materials between body cells and the sites of exchange; they must have appropriate support, coordination, and effector systems; and they must reproduce. Second, you know that different modes of life demand different means of solving the basic problems—an organism well suited to one sort of habitat may be ill suited to another. Every organism has mechanisms for coping with the peculiar problems its environment poses. Such mechanisms, encompassing structural, behavioral, physiological, and other characteristics, are called **adaptations**. Adaptations are not static. Environments change, and the sets of challenges that face their inhabitants are constantly undergoing modification. If the organisms are to survive, they, too, must change.

While individuals die, populations of organisms survive change through genetic variation, which, as we have repeatedly noted, has its origin in mutation and is extended and maintained by the recombination resulting from sexual reproduction. Should environmental change occur, variation within a population provides natural selection with an array of slightly different "problem solving" mechanisms upon which to act. Natural selection effects more rapid reproduction of individuals better fit to cope with environmental challenges and thereby encourages continual adjustment and "tuning" of organisms to shifting environmental demands.

In this laboratory, you will be asked to bring your understanding of adaptation to bear on the study of an unfamiliar organism.

Your instructor will select the "unknown" organism from among those that you have not yet studied in this course. Since you will not have time in one laboratory period to study every aspect of the organism's biology, you should limit your study to some area that particularly interests you. If you have a plant, for example, you might wish to compare its gas-exchange, supportive, vascular (if present), or reproductive systems with those of other plants you have studied; you might wish to compare (using chromatography) its pigments with those of other plants you have studied. If you have an animal, you might study its feeding behavior, locomotion, or other behavior that interests you; you might make a detailed structural comparison of some organ system (e.g., digestive system) with the same system in other animals you have already studied, and then make hypotheses about your animal's diet. Whatever study you choose, try to focus on adaptations unique to the organism you are studying, and try to deduce as much as possible about environmental pressures to which the organism is subject.

Any techniques of study (dissection, microscopy, staining, hand sectioning, etc.) available in your laboratory are open to you; your instructor may have specific techniques to suggest. Should you wish to undertake a project that involves materials not available in your laboratory room, make sure that your instructor knows of your plans and approves them.

Your instructor may provide references. Use them liberally as guides, but do not depend upon the observations of others in making your report. Raise questions, suggest answers to as many questions as you can, and be prepared to defend your hypotheses with concrete observations.

Turn in a report as outlined by your instructor. Detail your findings, and submit appropriate drawings.

APPENDICES

APPENDIX 1 DRAWING

There are many good reasons for making drawings in biology. Perhaps the most important one is that drawing forces you to observe closely and accurately. If you approach the necessity of drawing with intent to profit by using it as a tool to learn the techniques of close observation, you will never regret the effort expended.

Also, drawings form an important record of things you see. While photographs are limited to representing appearance in a single moment in time, drawings can be composites of the observer's cumulative experience with many different specimens of the same material. The total picture of an object thus represented can often communicate information much more effectively than a photograph, and communication is extremely important to scientists who wish to have their published works clearly understood by others.

The suggestions below will help you make effective drawings. If you follow the suggestions at the beginning of the course, they will soon become habitual.

1. Your drawing should be complete and an accurate representation of the material you have observed, and should communicate your understanding of the material to anyone who looks at it. Avoid making "idealized" drawings; your drawing should be what you actually see, not what you imagine should be there.

2. A drawing should be large enough to represent easily and without crowding all the details you see. Rarely, if ever, are drawings too large; but they are often much too small. Show only as much as is necessary for an understanding of the structure; a small section shown in detail will often suffice. It is time-consuming and unnecessary, for example, to reproduce accurately the entire contents of a microscope field.

3. Use of a 3H or 4H pencil, good, smooth-surfaced paper, and a clean eraser will enable you to produce maximum resolution with minimum effort.

4. Use only simple, narrow lines. Represent depth by stippling (dots close together). Indicate depth only when it is essential to your drawing. Usually it is not. The ratio of observations to lines drawn should be at least 10:1, not the reverse.

5. Leave plenty of room for labels. Label the drawing with (a) the source of the material; (b) magnification under which it was observed; (c) structures; and, in the case of living materials, (d) movements you have seen. Try to keep label lines parallel to one side of the drawing paper (usually the bottom), and avoid crossing label lines.

Drawings are not only records, but a means of encouraging close observation; artistic ability is unnecessary. Before you turn in a drawing, ask yourself if you know what every line represents. If the answer is negative, look more closely at the material you are working on.

APPENDIX 2   CLASSIFICATION

To deal with the vast array of life, biologists need an orderly system by which species can be classified in a meaningful manner. The classification system used in biology today consists of a hierarchy of graded taxonomic ranks. **Taxonomy** is the specialization within biology that deals exclusively with bringing order into the classification system. Each taxonomic rank automatically encodes information about every aspect of the organisms it includes, thereby conveying information useful in almost every other area of biology. Some of the ranks, in order of decreasing inclusiveness, are as follows:

Kingdom
  Phylum or division
    Class
      Order
        Family
          Genus
            Species

Each category in this hierarchy is a collective unit containing one or more groups from the next-lower level in the hierarchy. Thus, a genus is a group of closely related species unified by some characteristic considered more fundamental than the differences that define separate species of that genus. A family is a group of related genera, etc. Such a scheme represents an effort to group organisms according to relationship. Often relationships can be determined from fossils. But where fossils are not available, we are often forced to guess relationships by studying similarities in anatomy, development, behavior, and other characteristics of living organisms.

Because this system of hierarchy presumably reflects degree of kinship, we can set up "phylogenetic trees" to show probable evolutionary histories. Using the trees will help you to understand the hypothetical relationships.

The first time you encounter an organism, look up its genus in the adjoining table, and place its scientific name in the proper position on its phylogenetic tree. A few genera you use will not be found in the adjoining list. If you encounter such organisms and if their position is not obvious from context, ask your instructor to help. Each species is placed at the tip of a branch which you must draw on the tree yourself.

A CLASSIFICATION OF LIVING THINGS

KINGDOM MONERA

SECTION EUBACTERIA

DIVISION SPIROCHETES. *Spirochaeta, Treponema*
DIVISION CYANOBACTERIA. *Gloeocapsa, Microcystis, Oscillatoria, Nostoc, Scytonema*
DIVISION GRAM-POSITIVE BACTERIA
    CLASS CLOSTRIDIA. *Bacillus, Clostridium, Mycoplasma, Staphylococcus, Streptococcus*
    CLASS ACTINOMYCES. *Micrococcus, Actinomyces, Streptomyces*
DIVISION GREEN SULFUR BACTERIA. *Chlorobium*
DIVISION PURPLE BACTERIA
    CLASS PURPLE SULFUR BACTERIA. *Pseudomonas, Escherichia*
    CLASS RHODOPSEUDOMONAS. *Rhodopseudomonas, Rhizobium*
    CLASS PURPLE NONSULFUR. *Sphaerotilus, Alcaligenes*
    CLASS DESULFOVIBRIO. *Desulfovibrio*
DIVISION PROCHLOROPHYTA. *Prochloron*

SECTION ARCHAEBACTERIA

DIVISION METHANOGENS. *Methanobacterium*
DIVISION HALOPHILES. *Halobacterium*
DIVISION THERMOACIDOPHILES. *Thermoplasma*

KINGDOM PROTISTA

SECTION PROTOPHYTA: Algal protists

DIVISION EUGLENOPHYTA. Euglenoids. *Euglena, Eutreptia, Phacus, Colacium*
DIVISION CHRYSOPHYTA
    CLASS CHRYSOPHYCEAE. Golden-brown algae. *Chrysamoeba, Chromulina, Synura, Mallomonas*
    CLASS HAPTOPHYCEAE (or Prymnesiophyceae). Haptophytes and coccolithophores. *Isochrysis, Prymnesium, Phaeocystis, Coccolithus, Hymenomonas*
    CLASS XANTHOPHYCEAE. Yellow-green algae. *Botrydiopsis, Halosphaera, Tribonema, Botrydium*
    CLASS EUSTIGMATOPHYCEAE. Eustigmatophytes. *Pleurochloris, Visheria, Pseudocharaciopsis*
    CLASS CHLOROMONADOPHYCEAE. Chloromonads. *Gonyostomum, Reckertia*
    CLASS BACILLARIOPHYCEAE. Diatoms. *Pinnularia, Arachnoidiscus, Triceratium, Pleurosigma*
DIVISION PYRROPHYTA. Dinoflagellates. *Gonyaulax, Gymnodinium, Ceratium, Gloeodinium*
DIVISION CRYPTOPHYTA. Cryptomonads. *Cryptomonas, Chroomonas, Chilomonas, Hemiselmis*

SECTION PROTOMYCOTA: Fungal protists

DIVISION HYPHOCHYTRIDIOMYCOTA. Hypochytrids. *Rhizidiomyces*
DIVISION CHYTRIDIOMYCOTA. Chytrids. *Olpidium, Rhizophydium, Diplophlyctis, Cladochytrium*

SECTION GYMNOMYCOTA: Slime molds

DIVISION PLASMODIOPHOROMYCOTA. Plasmodiophores (or endoparasitic slime molds). *Plasmodiophora, Spongospora, Woronina*
DIVISION LABYRINTHULOMYCOTA. Net slime molds. *Labyrinthula*
DIVISION ACRASIOMYCOTA. Cellular slime molds. *Dictyostelium, Polysphondylium*
DIVISION MYXOMYCOTA. True slime molds. *Physarum, Hemitrichia, Stemonitis*

SECTION PROTOZOA: Animal-like protists

PHYLUM MASTIGOPHORA
    CLASS ZOOFLAGELLATA (or Zoomastigina). "Animal" flagellates. *Trypanosoma, Calonympha, Chilomonas, Trichonympha*
    CLASS OPALINATA. Opalinids. *Opalina, Zelleriella*
PHYLUM SARCODINA. Pseudopodal protozoans
    CLASS RHIZOPODEA. Naked and shelled amoebae, foraminiferans. *Amoeba, Pelomyxa, Entamoeba, Arcella, Globigerina, Textularia*
    CLASS ACTINOPODEA. Radiolarians, heliozoans, acantharians. *Aulacantha, Acanthometron, Actinosphaerium, Actinophrys*
PHYLUM SPOROZOA. Sporulation protozoans
    CLASS TELOSPOREA. *Monocystis, Gregarina, Eineria, Toxoplasma, Plasmodium*
    CLASS PIROPLASMEA. *Babesia, Theileria*
PHYLUM CNIDOSPORA. Cnidosporians
    CLASS MYXOSPOREA. *Myxobolus, Myxidium, Ceratomyxa*
    CLASS MICROSPOREA. *Nosema, Haplosporidium, Mrazekia*
PHYLUM CILIATA. Ciliates. *Paramecium, Stentor, Vorticella, Spirostomum*

KINGDOM PLANTAE

DIVISION CHLOROPHYTA. Green algae. *Chlamydomonas, Volvox, Ulothrix, Spirogyra, Oedogonium, Ulva*
DIVISION CHAROPHYTA. Stonewarts. *Chara, Nitella, Tolypella*

DIVISION PHAEOPHYTA. Brown algae. *Sargassum, Ectocarpus, Fucus, Laminaria*
DIVISION RHODOPHYTA. Red algae. *Nemalion, Polysiphonia, Dasya, Chondrus, Batrachospermum*
DIVISION BRYOPHYTA
    CLASS HEPATICAE. Liverworts. *Marchantia, Conocephalum, Riccia, Porella*
    CLASS ANTHOCEROTAE. Hornworts. *Anthoceros*
    CLASS MUSCI. Mosses. *Polytrichum, Sphagnum, Mnium*
DIVISION TRACHEOPHYTA. Vascular plants
    Subdivision Psilopsida. *Psilotum, Tmesipteris*
    Subdivision Lycopsida. Club mosses. *Lycopodium, Phylloglossum, Selaginella, Isoetes, Stylites*
    Subdivision Sphenopsida. Horsetails. *Equisetum*
    Subdivision Pteropsida. Ferns. *Polypodium, Osmunda, Dryopteris, Botrychium, Pteridium*
    Subdivision Spermopsida. Seed plants
        CLASS PTERIDOSPERMAE. Seed ferns. No living representatives
        CLASS CYCADAE. Cycads. *Zamia*
        CLASS GINKGOAE. *Ginkgo*
        CLASS CONIFERAE. Conifers. *Pinus, Tsuga, Taxus, Sequoia*
        CLASS GNETEAE. *Gnetum, Ephedra, Welwitschia*
        CLASS ANGIOSPERMAE. Flowering plants
            SUBCLASS DICOTYLEDONEAE. Dicots. *Magnolia, Quercus, Acer, Pisum, Taraxacum, Rosa, Chrysanthemum, Aster, Primula, Ligustrum, Ranunculus*
            SUBCLASS MONOCOTYLEDONEAE. Monocots. *Lilium, Tulipa, Poa, Elymus, Triticum, Zea, Ophyrys, Yucca, Sabal*

KINGDOM FUNGI

DIVISION OOMYCOTA. Water molds, white rusts, downy mildews. *Saprolegnia, Phytophthora, Albugo*
DIVISION ZYGOMYCOTA. Conjugation fungi
    CLASS ZYGOMYCETES. *Rhizopus, Mucor, Phycomyces, Choanephora, Entomophthora*
    CLASS TRICHOMYCETES. *Stachylina*
DIVISION ASCOMYCOTA. Sac fungi
    CLASS HEMIASCOMYCETES. Yeasts and their relatives. *Saccharomyces, Schizosaccharomyces, Endomyces, Eremascus, Taphrina*
    CLASS PLECTOMYCETES. Powdery mildews, fruit molds, etc. *Erysiphe, Podosphaera, Aspergillus, Penicillium, Ceratocystis*
    CLASS PYRENOMYCETES. *Sordaria, Neurospora, Chaetomium, Xylaria, Hypoxylon*
    CLASS DISCOMYCETES. *Sclerotinia, Trichoscyphella, Rhytisma, Xanthoria, Pyronema*
    CLASS LABOULBENIOMYCETES. *Herpomyces, Laboulbenia*
    CLASS LOCULOASCOMYCETES. *Cochliobolus, Pyrenophora, Leptosphaeria, Pleospora*

DIVISION BASIDIOMYCOTA. Club fungi
    CLASS HETEROBASIDIOMYCETES. Rusts and smuts. *Ustilago, Urocystis, Puccinia, Phragmidium, Melampsora*
    CLASS HOMOBASIDIOMYCETES. Toadstools, bracket fungi, mushrooms, puffballs, stinkhorns, etc. *Coprinus, Marasmius, Amanita, Agaricus, Lycoperdon, Phallus*

KINGDOM ANIMALIA

SUBKINGDOM PARAZOA

PHYLUM PORIFERA. Sponges
    CLASS CALCAREA. Calcareous (chalky) sponges. *Scypha, Leucosolenia, Sycon, Grantia*
    CLASS HEXACTINELLIDA. Glass sponges. *Euplectella, Hyalonema, Monoraphis*
    CLASS DEMOSPONGIAE. *Spongilla, Euspongia, Axinella*
    CLASS SCLEROSPONGIAE. Coralline sponges. *Ceratoporella, Stromatospongia*

SUBKINGDOM PHAGOCYTELLOZOA

PHYLUM PLACOZOA. *Trichoplax*

SUBKINGDOM EUMETAZOA

SECTION RADIATA

PHYLUM CNIDARIA (or Coelenterata)
    CLASS HYDROZOA. Hydrozoans. *Hydra, Obelia, Gonionemus, Physalia*
    CLASS CUBOZOA. Sea wasps. *Tripedalia*
    CLASS SCYPHOZOA. Jellyfish. *Aurelia, Pelagia, Cyanea*
    CLASS ANTHOZOA. Sea anemones and corals. *Metridium, Pennatula, Gorgonia, Astrangia*
PHYLUM CTENOPHORA. Comb jellies. *Pleurobrachia, Haeckelia*

SECTION PROTOSTOMIA

PHYLUM PLATYHELMINTHES. Flatworms
    CLASS TURBELLARIA. Free-living flatworms. *Planaria, Dugesia, Leptoplana*
    CLASS TREMATODA. Flukes. *Fasciola, Schistosoma, Prosthogonimus*
    CLASS CESTODA. Tapeworms. *Taenia, Dipylidium, Mesocestoides*
PHYLUM GNATHOSTOMULIDA. *Gnathostomula, Haplognathia*

PHYLUM NEMERTEA (or Rhynchocoela). Proboscis or ribbon worms
  CLASS ANOPLA. *Tubulanus, Cerebratulus*
  CLASS ENOLPA. *Amphiporus, Prostoma, Mala-cobdella*
PHYLUM ACANTHOCEPHALA. Spiny-headed worms. *Echinorhynchus, Gigantorhynchus*
PHYLUM MESOZOA
  CLASS RHOMBOZOA. *Dicyema, Pseudicyema, Conocyema*
  CLASS ORTHONECTIDA. *Rhopalura*
PHYLUM ROTIFERA. Rotifers. *Asplanchna, Hydatina, Rotaria*
PHYLUM GASTROTRICHA. *Chaetonotus, Macrodasys*
PHYLUM KINORHYNCHA (or Echinodera). *Echinoderes, Semnoderes*
PHYLUM NEMATA. Roundworms or nematodes. *Ascaris, Trichinella, Necator, Enterobius, Ancyclostoma, Heterodera*
PHYLUM NEMATOMORPHA. Horsehair worms. *Gordius, Paragordius, Nectonema*
PHYLUM ENTOPROCTA. *Urnatella, Loxosoma, Pedicellina*
PHYLUM LORICIFERA
PHYLUM PRIAPULIDA. *Priapulus, Halicryptus*
PHYLUM BRYOZOA (or Ectoprocta). Bryozoans, moss animals
  CLASS GYMNOLAEMATA. *Paludicella, Bugula*
  CLASS PHYLACTOLAEMATA. *Plumatella, Pectinatella*
  CLASS STENOLAEMATA
PHYLUM PHORONIDA. *Phoronis, Phoronopsis*
PHYLUM BRACHIOPODA. Lamp shells
  CLASS INARTICULATA. *Lingula, Glottidia, Discina*
  CLASS ARTICULATA. *Magellania, Neothyris, Terebratula*
PHYLUM MOLLUSCA. Molluscs
  CLASS CAUDOFOVEATA. *Chaetoderma*
  CLASS SOLENOGASTRES. Solenogasters. *Neomenia, Proneomenia*
  CLASS POLYPLACOPHORA. Chitons. *Chaeto-pleura, Ischnochiton, Lepidochiton, Amicula*
  CLASS MONOPLACOPHORA. *Neopilina*
  CLASS GASTROPODA. Snails and their allies (uni-valve molluscs). *Helix, Busycon, Crepidula, Haliotis, Littorina, Doris, Limax*
  CLASS SCAPHODA. Tusk shells. *Dentalium, Cadulus*
  CLASS BIVALVA. Bivalve mulluscs. *Mytilus, Ostrea, Ptecten, Mercenaria, Teredo, Tagelus, Unio, Anodonta*
  CLASS CEPHALOPODA. Squids, octopuses, etc. *Loligo, Octopus, Nautilus*
PHYLUM POGONOPHORA. Beard worms. *Siboglinum, Lamellisabella, Oligobrachia, Polybrachia*
PHYLUM SIPUNCULA. *Sipunculus, Phascolosoma, Dendrostomum*
PHYLUM ECHIURA
  CLASS ECHIUROINEA. *Echiurus, Ikedella*
  CLASS XENOPNEUSTA. *Urechis*
  CLASS HETEROMYOTA. *Crisia, Tubulipora*

PHYLUM ANNELIDA. Segmented worms
  CLASS POLYCHAETA (including Archiannelida). Sandworms, tubeworms, etc. *Nereis, Chaetopterus, Aphrodite, Diopatra, Arenicola, Hydroides, Sabella*
  CLASS OLIGOCHAETA. Earthworms and many freshwater annelids. *Tubifex, Enchytraeus, Lumbricus, Dendrobaena*
  CLASS HIRUDINOIDEA. Leeches. *Trachelobdella, Hirudo, Macrobdella, Haemadipsa*
PHYLUM ONYCHOPHORA. *Peripatus, Peripatopsis*
PHYLUM TARDIGRADA. Water bears. *Echiniscus, Macrobiotus*
PHYLUM ARTHROPODA
  Subphylum Trilobita. No living representatives
  Subphylum Chelicerata
    CLASS EURYPTERIDA. No living representatives
    CLASS MEROSTOMATA. Horseshoe crabs. *Limulus*
    CLASS ARACHNIDA. Spiders, ticks, mites, scorpions, whipscorpions, daddy longlegs, etc. *Archaearanea, Latrodectus, Argiope, Centruroides, Chelifer, Mastigoproctus, Phalangium, Ixodes*
    CLASS PYCNOGONIDA. Sea spiders. *Nymphon, Ascorhynchus*
  Subphylum Uniramia (or Mandibulata)
    CLASS CRUSTACEA. *Homarus, Cancer, Daphnia, Artemia, Cyclops, Balanus, Porcellio, Cambarus*
    CLASS CHILOPODA. Centipeds. *Scolopendra, Lithobius, Scutigera*
    CLASS DIPLOPODA. Millipeds. *Narceus, Apheloria, Polydesmus, Julus, Glomeris*
    CLASS PAUROPODA. *Pauropus*
    CLASS SYMPHYLA. *Scutigerella*
    CLASS INSECTA. Insects
      ORDER THYSANURA. Bristletails, silverfish, firebrats. *Machilis, Lepisma, Thermobia*
      ORDER EPHEMERIDA. Mayflies. *Hexagenia, Callibaetis, Ephemerella*
      ORDER ODONATA. Dragonflies, damselflies. *Archilestes, Lestes, Aeshna, Gomphus*
      ORDER ORTHOPTERA. Grasshoppers, crickets, etc. *Schistocerca, Romalea, Nemobius, Megaphasma*
      ORDER PHASMATOPTERA. Walking sticks. *Phyllium*
      ORDER BLATTARIA. Cockroaches. *Blatta, Periplaneta*
      ORDER MANTODEA. Mantids. *Mantis*
      ORDER GRYLLOBLATTARIA. *Grylloblatta*
      ORDER ISOPTERA. Termites. *Reticulitermes, Kalotermes, Zootermopsis, Nasutitermes*
      ORDER DERMAPTERA. Earwigs. *Labia, Forficula, Prolabia*
      ORDER EMBIIDINA (or Embiaria or Embioptera). *Oligotoma, Anisembia, Gynembia*
      ORDER PLECOPTERA. Stoneflies. *Isoperla, Taeniopteryx, Capnia, Perla*
      ORDER ZORAPTERA. *Zorotypus*
      ORDER PSOCOPTERA. Book lice.

Ectopsocus, Liposcelis, Trogium
ORDER MALLOPHAGA. Chewing lice.
*Cuclotogaster, Menacanthus, Menopon,
Trichodectes*
ORDER ANAPLURA. Sucking lice. *Pediculus,
Phthirius, Haematopinus*
ORDER THYSANOPTERA. Thrips. *Helio-
thrips, Frankliniella, Hercothrips*
ORDER HEMIPTERA. True bugs. *Belostoma,
Lygaeus, Notonecta, Cimex, Lygus, Oncopeltus*
ORDER HOMOPTERA. Cicadas, aphids, leaf-
hoppers, scale insects, etc. *Magicicada,
Circulifer, Psylla, Aphis, Saissetia*
ORDER NEUROPTERA. Dobsonflies, alder-
flies, lacewings, mantispids, snakeflies, etc.
*Corydalus, Hemerobius, Chrysopa, Mantispa,
Agulla*
ORDER COLEOPTERA. Beetles, weevil.
*Copris, Phyllophaga, Harpalus, Scolytus,
Malanotus, Cicindela, Dermestes, Photinus,
Coccinella, Tenebrio, Anthonomus,
Conotrachelus*
ORDER HYMENOPTERA. Wasps, bees, ants,
sawflies. *Cimbex, Vespa, Glypta, Scolia,
Bembix, Formica, Bombus, Apis*
ORDER STREPSIPTERA. Endoparasites
ORDER MECOPTERA. Scorpionflies.
Panorpa, Boreus, Bittacus
ORDER SIPHONAPTERA. Fleas. *Pulex,
Nosopsyllus, Xenopsylla, Ctenocephalides*
ORDER DIPTERA. True flies, mosquitoes.
*Aedes, Asilus, Sarcophaga, Anthomyia, Musca,
Chironomus, Tabanus, Tipula, Drosophila*
ORDER TRICHOPTERA. Caddisflies.
Limnephilus, Rhyacophilia, Hydropsyche
ORDER LEPIDOPTERA. Moths, butterflies.
*Tinea, Pyrausta, Malacosoma, Sphinx, Samia,
Bombyx, Heiliothis, Papilio, Lycaena*
Subphylum Pentastomida
CLASS PENTASTOMIDA. Parasites

SECTION DEUTEROSTOMIA

PHYLUM CHAETOGNATHA. Arrow worms. *Sagitta,
Spadella*
PHYLUM ECHINODERMATA
Subphylum Crinozoa
CLASS CRINOIDEA. Crinoids, sea lilies. *Antedon,
Ptilocrinus, Comactinia*
Subphylum Asterozoa
CLASS STELLEROIDEA. Sea stars, brittle stars.
*Asterias, Ctenodiscus, Luidia, Oreaster, Asteronyx,
Amphioplus, Ophiothrix, Ophioderma, Ophiura*
Subphylum Echinozoa
CLASS ECHINOIDEA. Sea urchins, sand dollars,
heart urchins. *Cidaris, Arbacia, Strongylocentrotus,
Echinanthus, Echinarachnius, Moira*
CLASS HOLOTHUROIDEA. Sea cucumbers.
Cucumaria, Thyone, Caudina, Synapa

PHYLUM HEMICHORDATA
CLASS ENTEROPNEUSTA. Acorn worms.
Saccoglossus, Balanoglossus, Glossobalanus
CLASS PTEROBRANCHIA. *Rhabdopleura,
Cephalodiscus*
CLASS PLANCTOSPHAEROIDEA
PHYLUM CHORDATA. Chordates
Subphylum Tunicata (or Urochordata). Tunicates
CLASS ASCIDIACEA. Ascidians or sea squirts.
Ciona, Clavelina, Molgula, Perophora
CLASS THALIACEA. *Pyrosoma, Salpa, Doliolum*
CLASS APPENDICULARIA. *Appendicularia,
Oikopleura, Fritillaria*
Subphylum Cephalochordata. Lancelets, amphioxus.
Branchiostoma, Asymmetron
Subphylum Vertebrata. Vertebrates
CLASS AGNATHA. Jawless fish. *Cephalaspis,**
Pteraspis,† *Petromyzon, Entosphenus, Myxine,
Eptatretus*
CLASS ACANTHODII. No living representatives
CLASS PLACODERMI. No living representatives
CLASS CHONDRICHTHYES. Cartilaginous fish,
including sharks and rays. *Squalus, Hyporion, Raja,
Chimaera*
CLASS OSTEICHTHYES. Bony fish
SUBCLASS SARCOPTERYGII
ORDER CERATODIFORMES. Australian
lungfish. *Neoceratodus*
ORDER LEPIDOSIRENIFORMES. Lungfish.
Protopterus, Lepidosiren
SUBCLASS ACTINOPTERYGII. Ray-finned
fish. *Amia, Cyprinus, Gadus, Perca, Salmo*
CLASS AMPHIBIA
ORDER ANURA. Frogs and toads. *Rana,
Hyla, Bufo*
ORDER CAUDATA (or Urodela). Salaman-
ders. *Necturus, Triturus, Plethodon,
Ambystoma*
ORDER GYMNOPHIONA (or Apoda).
Ichthyophis, Typhlonectes
CLASS REPTILIA
ORDER TESTUDINES. Turtles. *Chelydra,
Kinosternon, Clemmys, Terrapene*
ORDER RHYNCHOCEPHALIA. Tuatara.
Sphenodon
ORDER CROCODYLIA. Crocodiles and
alligators. *Crocodylus, Alligator*
ORDER LEPIDOSAURIA. Snakes and lizards.
*Iguana, Anolis, Sceloporus, Phrynosoma,
Natrix, Elaphe, Coluber, Thamnophis, Crotalus*
CLASS AVES. Birds. *Anas, Larus, Columba, Gallus,
Turdus, Dendroica, Sturnus, Passer, Melospiza*
CLASS MAMMALIA. Mammals
SUBCLASS PROTHOTHERIA
ORDER MONOTREMATA. Egg-laying
mammals. *Ornithorhynchus, Tachyglossus.*

*Extinct.
†Extinct.

SUBCLASS THERIA.   Marsupial and placental mammals.

    ORDER METATHERIA (or Marsupialia). Marsupials. *Didelphis, Sarcophilus, Notoryctes, Macropus*

    ORDER INSECTIVORA.   Insectivores (moles, shrews, etc.).   *Scalopus, Sorex, Erinaceus*

    ORDER DERMOPTERA.   Flying lemurs. *Galeopithecus*

    ORDER CHIROPTERA.   Bats. *Myotis, Eptesicus, Desmodus*

    ORDER PRIMATA.   Lemurs, monkeys, apes, humans. *Lemur, Tarsius, Cebus, Macacus, Cynocephalus, Pongo, Pan, Homo*

    ORDER EDENTATA.   Sloths, anteaters, armadillos. *Bradypus, Myrmecophagus, Dasypus*

    ORDER PHOLIDOTA.   Pangolin. *Manis*

    ORDER LAGOMORPHA.   Rabbits, hares, pikas. *Ochotona, Lepus, Sylvilagus, Oryctolagus*

    ORDER RODENTIA.   Rodents. *Sciurus,*
Marmota, Dipodomys, Microtus, Peromyscus, Rattus, Mus, Erethizon, Castor

    ORDER ODONTOCETA.   Toothed whales, dolphins, porpoises. *Delphinus, Phocaena, Monodon*

    ORDER MYSTICETA.   Baleen whales. *Balaena*

    ORDER CARNIVORA.   Carnivores. *Canis, Procyon, Ursus, Mustela, Mephitis, Felis, Hyaena, Eumetopias*

    ORDER TUBULIDENTATA.   Aardvark. *Orycteropus*

    ORDER PROBOSCIDEA.   Elephants. *Elephas, Loxodonta*

    ORDER HYRACOIEDEA.   Hyraxes, conies. *Procavia*

    ORDER SIRENIA.   Manatees. *Trichechus, Halicore*

    ORDER PERISSODACTYLA.   Odd-toed ungulates. *Equus, Tapirella, Tapirus, Rhinoceros*

    ORDER ARTIODACTYLA.   Even-toed ungulates. *Percari, Sus, Hippopotamus, Camelus, Cervus, Odocoileus, Giraffa, Bison, Ovis, Bos*

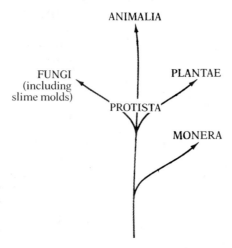

Fig. A2-1. The five-kingdom classification. Protista is recognized as a separate kingdom consisting of three groups derived from a common ancestor and leading to Plantae, Animalia, and Fungi.

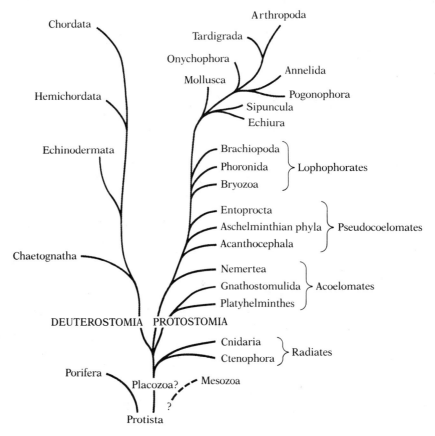

Fig. A2-2. One interpretation of evolutionary relationships among the animal phyla.

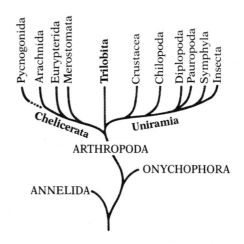

Fig. A2-3. Diagram of possible relationships among Annelida, Onychophora, and the major subphyla and classes of Arthropoda.

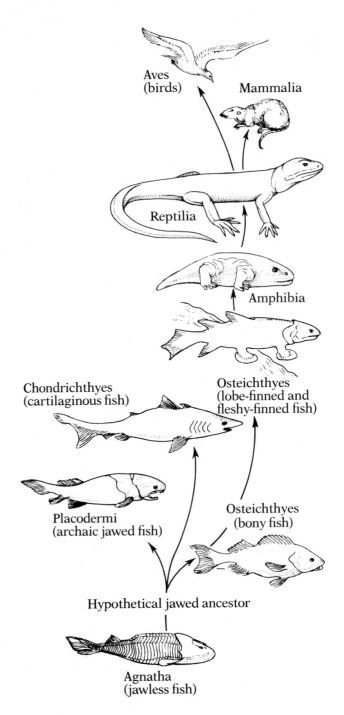

Fig. A2-4. Evolution of the vertebrate classes.

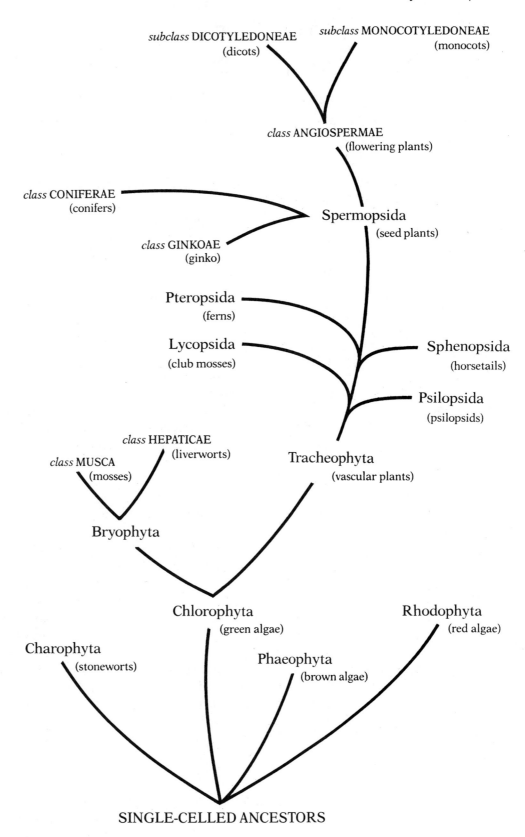

subclass DICOTYLEDONEAE (dicots)

subclass MONOCOTYLEDONEAE (monocots)

class ANGIOSPERMAE (flowering plants)

class CONIFERAE (conifers)

Spermopsida (seed plants)

class GINKOAE (ginko)

Pteropsida (ferns)

Lycopsida (club mosses)

Sphenopsida (horsetails)

Psilopsida (psilopsids)

class HEPATICAE (liverworts)

class MUSCA (mosses)

Tracheophyta (vascular plants)

Bryophyta

Chlorophyta (green algae)

Rhodophyta (red algae)

Charophyta (stoneworts)

Phaeophyta (brown algae)

SINGLE-CELLED ANCESTORS

PLANTAE

Fig. A2-5.

APPENDIX 3

PITHING AND DECEREBRATION OF A FROG

In many experiments with frogs, the central nervous system must be in-activated in order to eliminate sensation and response. This may be ac-complished by anesthesia; but most anesthetics wear off, and they may impair functioning of the internal organs. A relatively normal condition can be maintained for several hours with a simple operation known as **pithing**, or by **decerebration**.

PITHING

1. Obtain a frog that has been cooled by refrigeration or with crushed ice. Dry the animal on a paper towel.

2. Hold the frog in your fist, dorsal side up, with the head between your thumb and index finger. Now put your index finger on top of the head, and press down gently, bending the spinal cord at the neck.

3. With your free hand, bring the point of a dull* dissecting needle posteriorly down the middorsal line from the eyes until you feel the first indentation in the spinal column, just behind the skull. Set the dull needle aside, and, with a sharp dissecting needle, puncture the skin at this point, taking care to make a hole only in the skin.

4. Insert the dull needle into the spinal cord at this point, and push the needle forward, parallel to the external surface, as far as the needle will go. Move the needle to the right and left, destroying the brain. This *anterior* pithing destroys all of the brain but leaves the spinal cord intact. Complete destruction of the CNS entails pithing of the spinal cord, described in step 5.

5. Slowly bring the needle back to its point of entrance into the body. Without taking the needle out, turn it at a right angle to the body surface (pointing directly in), and then turn the handle back toward the head so that the pointed end, heading posteriorly, is again about parallel to the external surface. Gently push the needle into the spinal cord as far as it will go. If it has entered the spinal cord, the legs will jerk out straight. When you remove the needle, rotate it slowly so that all the nerves of the cord are disconnected.

*Dulled by a grindstone or on a cement floor.

318

DECEREBRATION

1. Obtain a frog that has been cooled by refrigeration or with crushed ice.

2. Use the anterior margin of the tympanic membranes (the brown circular membranes behind the eyes) as a guideline (Figure A3-1).

3. Hold the frog under cold running tap water. Insert a pair of heavy scissors into the mouth, and quickly cut off the anterior portion of the skull. Leave the lower jaw attached. You have removed the fore- and midbrain.

4. As in the procedure above, insert a dull needle into the spinal cord, and follow the instructions in "Pithing," step 5.

Fig. A3-1. Frog, showing cut to remove fore- and midbrain.